The Route of EDWIN WAY TEALE'S 19,000-mile Journey into Summer

JOURNEY INTO SUMMER

THE AMERICAN SEASONS

THE FIRST SEASON

NORTH WITH THE SPRING

THE SECOND SEASON

JOURNEY INTO SUMMER

THE THIRD SEASON

AUTUMN ACROSS AMERICA

JOURNEY INTO SUMMER

EDWIN WAY TEALE

A NATURALIST'S RECORD OF A 19,000-MILE JOURNEY THROUGH THE NORTH AMERICAN SUMMER, WITH PHOTOGRAPHS BY THE AUTHOR

DODD, MEAD & COMPANY · NEW YORK
1960

Dedicated to
DAVID
Who Traveled with
Us in Our Hearts

ACKNOWLEDGMENTS

FROM Fay H. Young, Captain of the *Resolute*, who carried us safely through the storm on the Lake of the Woods, to Mrs. Joseph A. Estabrooks, Postmistress of Hampton, Connecticut, who weighed a silver dollar, my thanks for varied help—before, during and after our summer journey—is extended to many persons. Some aided us in reaching places we wished to visit. Others read portions of the manuscript. Others supplied missing facts or helped in the identification or understanding of what we had seen. Still others offered the benefit of their specialized knowledge. I am indebted particularly to the following:

John W. Aldrich, Dean Amadon, Alfred M. Bailey, Gladys Baker, Charles M. Bogert, Rachel L. Carson, O. H. Clark, James Cope, William S. Creighton, Allan D. Cruickshank, Helen G. Cruickshank, Thomas C. Desmond, John Doerr, Frank Dufresne, Philip DuMont, Mrs. Joseph A. Estabrooks, V. C. Fishel, James Forbes, Irving Friedman, Donald V. Gray, Elizabeth C. Hall, Merrill Hammond, Walter Harding, Inez Haring, Cordia J. Henry, Bob Hines, Lockwood Jaynes, Clifford Keech, John Kieran, Oren Kimberly, Alexander B. Klots, Wendell Lamb, Russell Lee, Harold G. Male, Karl Maslowski, Harold F. Mayfield, Harold N. Moldenke, Stanley Muliak, Roy Muma, Robert J. Niedrach, H. H. Nininger, Harry C. Oberholser, John Pallister, Clint Paulson, George H. Peters, Roger T. Peterson, P. P. Pirone, Richard H. Pough, Harold W. Rickett, Herbert Ruckes, T. C. Schnierla, Herbert F. Schwarz,

ACKNOWLEDGMENTS

James A. Selby, Gertrude Selby, Roy W. Sheppard, Ralph P. Silliman, James Slater, Ken and Ada Slater, H. T. U. Smith, Marshall Sprague, James Thorp, Emma Toft, Olivia Traven, Asher Treat, A. H. Whittemore, Farida A. Wiley, M. Woodbridge Williams, Raymond D. Wood, Mary V. Wissler, Fay H. Young.

I am especially under obligation to Benjamin T. Richards, a friend of long standing, for copy-editing the manuscript, checking galleys and page proofs and preparing the index. His assistance has been invaluable.

Prior to book publication, the chapter on "Stone Dragonflies" appeared in *Audubon Magazine*. I wish to express my thanks to the editor for permission to include the material in this volume.

As on numerous occasions before, during more than two decades of close and cordial relationship, I am well aware of how deeply I am indebted to members of the Dodd, Mead staff in the editing and production of the book, particularly to Edward H. Dodd, Jr., Raymond T. Bond, S. Phelps Platt, Jr., John Blair and Ruby Carr.

Finally, I am sure that the reader will understand that my debt to my wife, Nellie, the most congenial companion for all seasons of the year, is greater than can be set down in words.

June 11, 1960
EDWIN WAY TEALE

CONTENTS

1 FRANCONIA SUNRISE—SHADOWS TO THE WEST   1
Mountain shadows—A curious illusion—Summer's dawn—The constancy of the seasons—A wayfarer's summer—Our adventure with a season begins.

2 WALKING DOWN A RIVER   6
Naturalists and pedestrians—The snail's pace—Sunday River—The water road—Singing birds—The music of falling water—Track of a wildcat—The cranberry tree—A wilderness apple orchard—The curious doe—Veery songs—Night on the river.

3 SMUGGLER'S NOTCH   14
Bobolinks and blue flags—The rivers of sap—Smuggler's Notch—Ferns—A pioneer collector—The green spleenwort—Foam flowers—Face of the mountain—Foot Path in the Wilderness—Cottongrass—Painted trillium—The crown of clouds—Flies on a mountaintop—The Noise Needers—Thrush chorus—"Songster of the Woods"—The green days—We travel west.

4 NATURAL HISTORY OF NIAGARA FALLS   25
Niagara lights—Birds on the brink of the falls—The soundless sounds of Niagara—Birds riding the rapids—Waterlogged gulls—Rain that never stops—The silky alga—Pigeons and waterfalls—The dual character of the pigeon—Rainbows—The child's delight—A botanist at Niagara—The nonchalant mallard.

5 MAYFLY ISLAND   36
A mayfly storm—Effects of the invasion—Odd beliefs—The hidden life of the mayfly—Lights of a Lake Erie

x　　　　　CONTENTS

amusement park—The island of mayflies—Endurance of the weak—Luminous wings—Unique transformation—Husks of life—Amazing chitin—The charm of woodcuts—Sound of the mayfly swarm—W. H. Hudson's "fairy-dance of life"—Plunge of the dancers—The magnetism of light—Origin of this attraction—The darkened ferry.

6 THE MICELESS HOUSE　　　　　50

The Green Men—Summer leaves—Mowing—Source of the perfume in new-mown hay—Why grass is not killed by mowing—The dispossessed mice—A hundred miles of mouse paths—The ground sloth's contemporary—Fertility—New-born mice and water—The "mouseometer"—Green pastureland—Goatsbeard—Mouse appetites—Puzzling piles of timothy stems—Cycles of abundance—The Miceless House.

7 EIGHTY-FIVE MILES TO
　BREAKFAST　　　　　60

Lake country—Rifle River—Strawberry pie—The three-toed woodpecker—A wayfarer's breakfast—New words—Twinflowers—The first lake—Passenger pigeons—Ring around a lake—Across invisible boundary lines—The growing season—Passing of the horse—An Indian naturalist—New page on the calendar.

8 RIVER OF THE FIREFLIES　　　　　70

La Salle's portage—The great Kankakee marsh—River of 2,000 bends—The Carolina paroquet—Meat-hunters—Our zigzag course—Mint farms—The swelling humus—Bur oak leaves—Luminous animals—Trees decorated with fireflies—Nameless roads—Temperature and the flashing of fireflies—Source of the light—Rhythms—Synchronized flashing in fireflies—A great display—Cows surrounded by fireflies—A firefly chant of the Ojibway Indians.

9 WOODPECKER BLOCKADE　　　　　84

The red-headed woodpecker—The blockade—Resourcefulness of starlings—Varieties of approach—A shift in tactics—Victory by stealth—A conflict of instincts—The story of two robins—Birds feeding nestlings of a different species—The three baby flickers—Instinctive abandoning of the nesting hole—Corroborative evidence—What does *he* do?

CONTENTS

10 THE ORCHID RIDGES   95

Horicon marshes—Rails and gallinules—A celebrity—Kettle moraines—The Door Peninsula—A paradox of climates—Waves of sand—The Orchid Ridges—Rabbits and lady's slippers—A life spent among wildflowers—The perfume of the orchid—Deer and scented flowers—Beauty of a fawn—The courage of the wild—A crippled martin—The blind starling—A squirrel on the White House lawn—Compensating for handicaps—Stratagem of an eagle—Through the arch of a storm—Gully-washer.

11 HIGH ROCKS   108

Bracken—Leatherleaf—Wild Rice pancakes—Sugarbush Hill—A mountain-climbing botanist—His long hobby—Bear wallows—Granite Peak—The high rocks—Changing records—Sky pilot—The Wisconsin dusk.

12 WINGS IN THE SUN   119

Wild Rose Point—Sleepers and weekers—Ghost insects—The sleeping flies—Cornish pasty—Chipmunks—Canada geese—Bunny-in-the-grass—The eagle's nest—Birds struck by lightning—Yellow rails—Shining wings—Eclipse plumage—Tolling a loon—The ancient voice—Sandhill cranes—Swamp candles—Flight of the cranes—Whippoorwills—Days stolen from time.

13 MYSTERIOUS MAPLES   130

The Red Queen—Claw marks—The bird's-eye maple—Mystery trees—The timber cruiser—Riddle of the bird's-eyes—The many theories—The man with the destructive toe—A room with many eyes.

14 THE DARK RIVER   137

Stairway of the lakes—The largest lake—The wild Upper Peninsula—Hiawatha country—The gulls of Grand Portal—The nervous man of Miner's Castle—Northern swamp—Dark river—A wilderness waterfall—Witness trees—Otter slides—The sea lamprey—Mosquitoes—Wild strawberries.

15 BEARS OF COPPER HARBOR   150

The happy men—Copper Harbor—Ancient mines—A waterfall cuts through a vein of copper—Playground of the ravens—Bear pits—The watcher of bears—An ursine pecking order—Differences in three bears—A nose for honey—Mayfly titbits—The Porcupine Mountains—The year in its highest beauty—Skipping stones.

CONTENTS

16 FARTHEST NORTH   159
Raven roads—Red water—We leave the Great Lakes behind—Beaverwood—"Dazzle of the poplars"—Lake of the Woods—The *Resolute*—Storm on the Big Traverse—Disappearance of the sturgeon—Northwest Angle—A river from Manitoba—Wild rice shallows—The three-day summer—Waterlilies—Baby kingbirds breaking home ties—Quivering wings—The hunger movement transformed into flight—Instinctive transitions—Rain bubbles—Lost men—Northern fireflies—The frogs of the river shallows.

17 THE WATER PRAIRIES   170
Midsummer on the Great Plains—Souris, the river of mice—Land of lost lakes—Artesian water—Dragonflies keep pace with an automobile—Badgers—Virgin sod—Longspurs—The buffalo rubbing rock—Upland plover—The meeting of migrants—Godwits—Phalaropes—White pelicans—The baby grebe—Dance of the midges—The fecundity of summer.

18 BETWEEN HAY AND GRASS   179
Prairie dawn—Plentywood—Heat waves—Wheat—The silent birds—Heat of the Yellowstone—Lark buntings—A graveyard of dinosaurs—Bison land—Tongue hunters and hide hunters—"Mormon signposts"—Powder River—The noble huntsman—Sage hens—The colored mirage—Dreams in color—A time between hay and grass—Twin rainbows.

19 HOME OF THE PRAIRIE DOGS   188
Wishtonwish—Prairie dogs around us—Sociability—Play and work of the prairie dogs—Voice—Territorial bark—Dislike of rain—Nose prints—Dikes—Odd behavior of a prairie dog—Plunge holes—Complexity of the burrows—An enduring fable—Marshmallows and bubble gum—Cowboy's delight—Grasshopper hunters—War of extermination—Prairie dogs and water.

20 DRY RAIN   201
Eye's reach—Crossroads—Coneflowers—Spearfish Canyon—Flesh flies in the Black Hills—Fringes of a dust storm—The air-borne soil—Dust from a wet sandbar—Yellow light—Windbreaks—The golden earth—Source of Nebraska's loess—Dust sorted by the wind—Sandblast—Invisible Wayside—Static electricity—Sunset in a dust storm—A land of green dunes.

CONTENTS

21 GREEN DUNES   210

The contrary sheep—Individuality of the seasons—Hay meadows—Howlers and whitecaps—The underlying sand—An active snake—The great gulf—Road of the potash towns—A grouse tragedy—Dodder—One-cloud rains—Evaporating showers—Scotts Bluff—Pioneer wheeltracks—Why birds dive in front of speeding cars—Across America on two cylinders—We arrive at the average elevation of the continent.

22 GRASSHOPPER ROAD   221

Road paved with grasshoppers—How insects meet the summer heat—Desert leafhoppers—Inability of young birds to control their temperature—The lizard's sun—Lighter plumage on summer birds—Saliva saves the life of the kangaroo rat—Grasshopper wax—An insulated beetle—Estivation—Fence-post shadows—Shaded jackrabbits—Death Valley ants—We wear the bonnet of a cloud.

23 THE CORN WIND   230

The slack-water time of summer—A bridge without a river—Beauty of agricultural fields—Corn in the breeze—Shades of green in nature—Hybrids—The corn wind—Growing centers—A thirty-foot cornstalk—Design of stalk and leaf—Auctioneers—Watermelon sherbet—Gay-feathers—Goldfinch and bluebird—The state birds—Memories of a Model T.

24 TEN THOUSAND STEPS   240

The streamside willow—Highway of a river bed—*The Seasons*—Lined lizard—The house with yellow eyes—The Arkansas River—Floodwrack—Importance of the willow—Battle between ants and a ground beetle—Recollections of a highwayman—The heat wave breaks—Summer rain—The joy of the toads.

25 NIGHT OF THE FALLING STARS   251

Prairie dark—Meteor show—The flaming trails—A Kansas meteorite farm—Black rocks—Mrs. Kimberly's collection—A farmhand's find—The thrown-away meteorite—Meteor Crater—"Nuisance spots"—Meteor trails, blue and yellow—The meteor hunter—Thunder stones—Moon shadows—A kitten eats graham crackers—The fourth shadow—Blue meteor—brass sun.

xiv   CONTENTS

26 THE GLASS MOUNTAINS   264
Giant well—Snow-on-the-mountain—Valley of the butterflies—Scissor-tailed flycatchers—The curious song of a cicada—Mississippi kites—Greatest heat of the trip—The long sleep—Snowy plover—The Great Salt Plains—Upsidedown river—The Glass Mountains.

27 ON THE RIM OF THE WORLD   273
Stone fenceposts—A census of trees—The Osage orange—Riddles of distribution—Filling station sparrows—Hit by a grasshopper—The dwarfing of the sunflowers—Level land—Shrikes—Roads without turnings—A one-field dust storm—Cyclones and dust devils—Rain makers—Memories of our third home.

28 PARADE OF THE DUSTY TURTLES   284
Nest-filled trees—The plume Eriogonum—Chili coyote—The ornate box turtle—Its abundance—Habits of a dry-country turtle—Time sense—The "leaping turtles"—Avoidance of water—A turtle's remarkable method of removing ants—The Texas plain—Mustang country—Daniel Webster on the West—Yuccas—Jackrabbit ears—Coronado's trail—Pass of the pack rats—Goggles on a cow pony—The sea of sunflowers.

29 STONE DRAGONFLIES   294
The valley of fossil insects—Volcanic ash—White trees of stone—Victims of an insect Pompeii—Richness of the fossil beds—Fossil fever—A rosebud 10,000,000 years old—Leaf of a prehistoric water elm—"Devices of the Devil"—The pathetic Beringer—Pigmy nuthatches—Two remarkable fossil hunters—We find an ancient cranefly.

30 HIGH TUNDRA   304
Alpine meadows—Tundra in the sky—Bistort—The snow-lover—Alpine flora of the Rockies—Snow willow—Mountaintop soil—Little red elephant flower—Snowfield and rock slide—A sphinx moth visits a columbine—Insects from the sky—Grasshopper glaciers—Rosy finches—A marsh hawk two miles high—Mountain weathers—Wind at timberline—Pipits—Upper limit of trees—The sun god—Woolly stems—Earth cores of the pocket gophers—Marmots—Stillness of the heights.

CONTENTS

31 SUMMER SNOW   315

Grasshopper Park—The Gunnison country—Friends in a summer cabin—Pack rats—The mystery of spiral trees—Oh-Be-Joyful Creek—Migration in the mountains—The high road of the hummingbirds—Altitudinal migration in the Rockies—The "walking mountain"—Descending sheep—Signs of approaching autumn—Ground squirrels among the sunflowers—Colorado peaches—The million dollar wind—Black Canyon of the Gunnison—The poorwill—A lichen-covered fireplace—Unnamed peaks—Emerald Lake—Perennial snow—The snowline—Our fortunate accident—Stranded near home.

32 THE SAND BANNERS   325

The Great Sand Dunes—Migrating hawks—The stairway of the eagles—Banners of sand—Changing colors of the dunes—Angle of repose—Voice of the dunes—Lemonweed—The sagacious caterpillar—A double tree-line—Cloud Niagara—Kangaroo rats—Flashlight hunting—Web-footed horses.

33 LAND OF THE FIVE SUMMERS   333

The Rockies by air—Life zones—Colorado's five summers—Stubble ducks—Rattlesnake grass—The concentrated sweetness of summer's end—Two seasons in a valley—Blue spruce—Goldenweed—Human migration—Leaf fall—Mesa Verde—Pinyon jays—The Grand Hogback—Book Cliffs—The "city in the clouds"—A stairway of beaver ponds—Gold country—Wind timber—Pike's Peak.

34 THE GREAT EYRIE—SHADOWS TO THE EAST   342

Snow in the night—Blocked road to the peak—The highway is opened—We start up—Twenty-mile climb—Boulder-fields—Our highest point—Traffic congestion—Mountaintop visitors—The far view—America's heritage of beauty—Two businessmen—The clubs of the world—Shadow of the peak—Its advance across the plain—The last shadow of summer touches the horizon—The darkening land—The twilight of autumn's eve—Descent into the night—The fireplace—Red aurora—The mountain and the stars—Summer's end.

INDEX   349

ILLUSTRATIONS

Sunday River	facing page	12
Robin at nest	” ”	13
Painted trillium	” ”	44
Violets, foam flowers and ferns	” ”	45
The green spleenwort	” ”	45
Mayflies on a tree trunk	” ”	76
Barn swallows ready to leave the nest	” ”	77
The last swallow to leave	” ”	77
The Kankakee River in summer	” ”	108
George H. Peters and his high rocks	” ”	109
Firefly on a grass head	” ”	109
Fawn	” ”	140
Northern bog at sunset	” ”	141
Showy lady's slipper	” ”	172
Red squirrel	” ”	173
Least chipmunk	” ”	173
Nellie	” ”	204
Rushes at the Souris refuge	” ”	205
Dry stream in Montana	” ”	205
Erosion gully in dry country	” ”	205
Thunderhead over a wheat field	” ”	236
The rim of the world	” ”	237
Buffalo rubbing rock	” ”	237
Prairie dogs under Devils Tower	” ”	252
Visiting prairie dogs	following page	252
A prairie dog eating	” ”	252

ILLUSTRATIONS

Alert prairie dog	following page	252
Sitting beside the burrow entrance	" "	252
Emerging into the summer sunshine	facing page	253
Willow beside the Arkansas River	" "	268
Mud flats near the Great Salt Plains	" "	269
Cactus	" "	269
Above timberline in the Rockies	" "	284
Mountain meadow	following page	284
Alpine goldflower	" "	284
Arctic gentian	" "	284
Queen's crown	" "	284
Yellow paintbrush	" "	284
Bistort	" "	284
Colorado columbine	" "	284
Rocks in a mountain meadow	" "	284
Florissant shale bed	facing page	285
Fossil leaf of prehistoric water elm	" "	285
Fossil cranefly from Florissant	" "	285
Engelmann spruce near treeline	" "	316
Gunnison country near Emerald Lake	following page	316
Timberline tree	facing page	317
Storm-wracked spruce at timberline	" "	317
Spiral grain in dead tree	" "	317
Lemonweed among the Great Sand Dunes	" "	332
Stream disappearing at the edge of the dunes	" "	332
Shadow of Pike's Peak	" "	333
The shadow reaches the horizon	" "	333

JOURNEY INTO SUMMER

ONE

FRANCONIA SUNRISE—SHADOWS TO THE WEST

DOWN the long drop of the mountainside Lafayette Brook trailed the white thread of its foaming waters. Its headlong descent carried it seventy feet beneath the green bridge rail against which we leaned in the June dawn. Below us forested slopes fell away in a long toboggan out onto the valley floor. Behind us the Franconia Mountains rose darkly, hunching shoulders of stone against the sunrise, flinging an immense shadow mile after mile over the outspread land below.

As we watched, the farther edge of this shadow crept stealthily toward us. No direct movement caught our eye. But each time we glanced back more of the valley lay in sunshine. Over the roads and pastures, over the barns and houses, over Gale River and Meadow Brook, the narrowing shade trailed its farther edge. The invisible sun behind us rose; the visible shadow before us shrank. The effect was a curious illusion. Dawn and the light of the day seemed advancing toward us not out of the familiar east but out of the foreign west.

Here in this same spot, on this same bridge, just north of Franconia Notch in the White Mountains of New Hampshire, we had leaned against this same rail at the end of our long journey north with the spring. Here we had watched, over the valley, the flames of sunset ebbing slowly away. That was the sundown; this was the dawn. That was the end of a journey with a season; this was the beginning of another. Here, where we had bade farewell to the last sunset of spring, we were standing in the initial sunrise of summer.

Ten years separated those two events. Yet such is the constancy of the recurring seasons that a single period of starlight seemed to have intervened. Ten years before, a hundred years hence, the same bird songs would come down from these mountain slopes at dawn. The seasons, like the rivers of Ecclesiastes, endlessly return to their beginnings again. We were taking up where we had left off a decade before. Breathing in the perfumed air of the mountain dawn, Nellie and I might have been standing in the very daybreak that had followed. Summer was coming at its appointed time and we had come to meet it.

Our car, this time blue and white, stood waiting beside the road. It was packed with cameras and binoculars, journal paper and field guides, tramping shoes and raincoats. The accumulated pencil stubs of recent years filled a brown paper bag. Extra film and extra lenses were securely stowed away. And in the trunk more than 100 three-by-five-inch spiral-ring notebooks were housed in compact little boxes, a dozen to a box. Our maps—with contour quadrangles for special areas—were marked. Our plans were made. For thousands of miles, through more than half the states of the Union, from the backbone of New England—the White Mountains—to the backbone of the continent—the Rockies—my wife and I would wander as we chose in a winding, wayfarer's summer. All the season—three months, one day, fifteen hours and six minutes long—stretched away before us.

Spring and autumn are constantly changing, active seasons. Summer is more stable, more predictable. We tend to consider it the high point of the year, with spring moving toward it and autumn retreating from it. In summer life is easier, food and warmth more abundant. Babies born then have a lower infant mortality rate than at any other season of the year. When, some years ago, Columbia University psychologists conducted a survey, they found that, other factors being equal, most persons have the highest level of good feeling, the greatest sense of well-being, in the summertime.

To the average person, summer is the friend, winter the enemy. The Twilight of the Gods, in the old Norse legend, came with years that had no summers. Instinctively summer is accepted as the normal condition of the earth, winter as the abnormal. Summer is "the way it should be." It is as though our minds subconsciously returned to some tropical beginning, some summer-filled Garden of Eden.

As we stood on the bridge watching these shadows of the first summer day pull back and disappear around us, listening to whitethroats and wood thrushes in the dawn, watching a red squirrel gathering maple keys from the topmost twigs below us, the coming of the second season was ushering in the longest day of the year. At the moment of the season's beginning, at the summer solstice, the sun looks down from its northernmost point in the sky. The farthest boundary of its apparent northern swing is the Tropic of Cancer, which runs south of the tip of Florida. Thus never—in spite of appearances to the contrary—is the sun directly overhead anywhere, at any time, in the continental United States. On this twenty-first of June, the day of the summer solstice, the rays of the sun shone vertically down on a point in the Atlantic Ocean 360 miles north of the Virgin Islands and about 700 miles east and south of Florida's southern tip. Here for an instant the sun was "standing still in the sky" before it began what appeared to be its slow journey back to the Equator and the ending of summer.

Between these two events in time and space stretches the season of warmth and sunshine. Summer is vacation time, sweet clover time, swing and see-saw time, watermelon time, swimming and picnic and camping and Fourth-of-July time. This is the season of gardens and flowers, of haying and threshing. Summer is the period when birds have fewer feathers and furbearers have fewer hairs in their pelts. Through it runs the singing of insects, the sweetness of ripened fruit, the perfume of unnumbered blooms. It is a time of lambs and colts, kittens and puppies, a time to grow in. It is fishing time, canoeing time, baseball time. It is, for millions of Americans, "the good old summertime."

But America has many summers. Its continental span embraces the summer of the shore, the summer of the forest, the summer of the Great Plains, the summer of the mountains. We had chosen our general route to carry us through the greatest variety. We would see the season in vacation spots —along lake shores, on the mountain heights, in the cool north woods. But we would also see it on salt flats and in corn country, amid swamps and in areas where falling rain would be sucked up by the thirsty atmosphere before it reached the ground.

Our wandering course into the heart of the land would lead us through this diversity of America's second season. And along the way, if we were fortunate, we would observe such things as the winged whirlwind of a mayfly storm, prairie dogs at home, meteors streaming over the Great Plains in an August Perseid shower, the grace of Mississippi kites tilting in the wind and the high timberline meadows of the Rockies with their Alpine flowers still in bloom.

We had awakened that morning in the same small cabin beside the Pemigewasset River where we had sat before a fire of blazing logs that last night of spring a decade before. On a previous day we had left home amid the dawn chorus of the robins, crossed from Long Island and ridden north beside the Connecticut River through shimmering, hazy heat. Now the scented air of the mountains was dawn-cool.

When at last we climbed into our loaded car, sunshine filled all the valley. During those first of the summer hours we drove leisurely northward, along the vast sweep of land to the west of the Cherry Mountain Road, between the Shelburne birches, over the New Hampshire line and into Maine. There we turned north from Bethel and stopped at last beside wild Sunday River, tumbling out of the mountains with veeries singing in the forests of its banks.

Thus on a golden day in June we set out together. Once more we were adventuring across the months of an American season. We were at the beginning of what the old-time writers would have called "our joyful travels" through the many summers of the land.

TWO

WALKING DOWN A RIVER

EDEN PHILLPOTTS, the English author who has spent most of his life writing about lonely Dartmoor, tells of meeting a stranger at a luncheon in London. The man asked him if he knew Dartmoor.

"Not as well as I could wish," Phillpotts answered.

It developed that his companion had once motored across the sparsely settled tableland.

"Hardly a thing you would do twice, certainly."

Phillpotts suggested the next time he try walking across it. The man was greatly amused by this advice.

"Surely," he said, "no sane man would waste his time like that."

Yet Phillpotts knew that the way to become acquainted with an area intimately, to appreciate it best, is to walk over it. And the slower the walk the better. For a naturalist, the most productive pace is a snail's pace. A large part of his walk is often spent standing still. A mile an hour may well be fast enough. For his goal is different from that of the pedestrian. It is not how far he goes that counts; it is not how fast he goes; it is how much he sees.

And, in deeper truth, it is not *just* how much he sees. It is how much he appreciates, how much he feels. Nature affects our minds as light affects the photographic emulsion on a film. Some films are more sensitive than others; some minds are more receptive. To one observer a thing means much; to another the same thing means almost nothing. As the poet William Blake wrote in one of his letters: "The Tree which moves some to tears of joy is in the Eyes of others only a Green thing that stands in the way."

Under the morning sun of the second day of summer, Nellie and I began walking down wild Sunday River. A side road, first hardtop, then gravel, then dirt, then, at last, little more than wheeltracks leading on, had carried us to the upper reaches of the tumbling stream.

As we changed into rubber-soled sneakers that would cling to rocks and let us wade through shallows, we looked about us. On all sides, green-clad mountains gazed down upon us. Hardly four miles to the west the Appalachian Trail swung sharply north toward Goose Eye and Old Speck mountains. Eight miles or so to the east flowed the beautiful Androscoggin. And only twenty miles to the north Lake Umbagog straddled the Maine–New Hampshire line. There, three-quarters of a century before, William Brewster, the Cambridge ornithologist, had floated in his canoe, notebook on knee, making observations for his pioneer study of the avian life of one small area, *The Birds of the Lake Umbagog Region of Maine.*

To the lost man, to the pioneer penetrating new country, to the naturalist who wishes to see wild land at its wildest, the advice is always the same—follow a stream. The river is the original forest highway. It is nature's own Wilderness Road.

Long past now were the torrents of spring, and Sunday River rushed and gurgled down only half its rocky bed. Without haste, we followed its course as it flowed, plunging down little waterfalls, cascading among granite boulders, sliding over tilting ledges, fanning out into quiet pools. Once the whole river fell over a brink in a plunge of foaming water. Again the stream slowed into long, lazy stretches sparkling in the sun. We advanced mainly from rock to rock, soon learning to avoid wet patches of moss, as slippery as ice. Steppingstones down a river—this was the story of much of our travels that day. We progressed slowly, stopping often. We had no schedule. We had no goal except our own enjoyment of this alder-bordered north-country stream flowing through a land of singing birds.

They were everywhere around us, these birds of the forest watercourse—parula and Nashville and chestnut-sided warblers, redstarts and yellowthroats, chickadees and phoebes, blue jays sounding a raucous alarm, and robins, a few still nesting, singing in the higher trees. Always with us was the monologue of the red-eyed vireo and the endlessly repeated "Chebec!" of the tireless least flycatcher. Bluebirds called over us, and once a flock of goldfinches went roller-coastering by above the treetops. Time after time, clear above the rush of the stream, we heard the ovenbird set the forests ringing. But predominant among all this music of early summer there was that sweet, minor, infinitely moving lament, that voice of the north woods, the unhurried song of the little white-throated sparrow.

Somewhere, far up this forest stream, we came upon an old farmhouse—low, lonely, huddled in a tiny valley within sound of the water. Under its roof, living in these primitive surroundings, Kenneth Roberts, twenty years before, had written some of the finest pages of his *Northwest Passage*.

Half a mile or so downstream, I remember, we sat for a long time beside a diminutive waterfall only a foot or two high, delighting in the low music that filled the air. The water gurgled and hissed, lisped and murmured. Never before had we appreciated quite so clearly how many rushing, bubbling, liquid sounds combine to form the music of falling water. All down the river, all through the day, all through the night, the song of the running water went on and on. Light and dark are the same to a flowing stream. For it, only gravity matters.

Where a shelving rock of gray granite slanted down above another little waterfall, we ate our lunch that noon, a lunch of buns and cheese, small ripe tomatoes and sweet red grapes. Our extra dessert was the finest imaginable, handfuls of wintergreen berries picked on the mossy slope of the bank above. As we sat there little keys of moosewood came whirling down around us. Across the river, a redstart, a parula and a magnolia warbler sang, the parula repeating over and over its whirling upward trill that concluded in a kind of trip or turnover at the end; the magnolia, like the redstart, short and explosive, but with notes more rounded and whistled. Above the lower trees of this farther bank, an ash leaned out over the stream. From its dead top, a cedar waxwing over and over darted out, snapped up a flying insect and returned to its perch again. Its curving course cut through the air like the swing of a scythe. It was engaged in its own aerial mowing, gathering in its own harvest of summer.

Working downstream that afternoon, we came to turns where all the young birches and alders were tilted in the same direction, bent by the high waters of spring. At times we skirted rapids, pushing our way among the streamside alders. Again we clambered over great tangles of driftwood, floodwrack caught among the trees and bushes.

Up and down the riversides, back and forth across the water, butterflies came drifting by that day. They were mainly tiger swallowtails, here the northern subspecies, *canadensis*, slightly smaller and somewhat darker than the more southern forms. In the eighteenth century, when Oliver Goldsmith was writing his natural history and came to the butterflies, he noted: "Linnaeus has described near 800 kinds of this beautiful insect; and even his catalogue is allowed to be incomplete." Once when I was talking with Dr. Alexander B. Klots—whose *A Field Guide to the Butterflies* is now the standard work for eastern North America—I asked him how incomplete that original list really is. For the United States, just east of the Great Plains, he told me, the number of species and subspecies now far exceeds Linnaeus' figure. For the whole world, his 800 might be multiplied a hundred times and still fall short of the total. Those simple days of unawareness when Goldsmith could say there are "three or four kinds" of dragonflies and John Josselyn, Gent., writing of early New England, could speak of "the wasp," those days have long since gone. The insect hosts have not increased, but our awareness has.

At times a remembrance will be hard and solid like an artifact, a piece of pottery or an arrowhead found in the sand. At other times a memory will be quicksilver. We can never quite grasp it or hold it. It slides through our fingers when we try to squeeze it tight. One recollection of our wandering along the river comes back as clear-cut as the imprint we found in the hardening mud of a streamside puddle. Here a wildcat had set its foot, leaving behind a track as delicately defined as a plaster mold.

At the top of the bank nearby, above the green mats of the bunchberry starred with white, a cranberry tree spread its broad, three-pointed leaves. This north-country viburnum, *Viburnum opulus*, is variously known as the squaw bush, the water elder, the high-bush cranberry and the pincushion tree. Across its top, like clots of river foam, ran the white masses of its flowers. In the autumn they would be replaced by the red fruit, sour and translucent, that provided the pioneers with a substitute for cranberries and a jelly that, while tart, was less bitter than that of the related wayfaring tree. Each fall, like the pioneers of old, the ruffed grouse harvest the acid berries.

Past undercut forest floor, past rocks overlaid with moss, past shaded banks shaggy with reindeer lichen, past the slide of water down some long chute of tilted rock we followed the stream in its windings. In many places we stopped to examine depressions like bowls or basins worn into the ledges and larger rocks. Each contained smooth, rounded pieces of granite. They ranged from pebbles to balls the size of grapefruit. The larger the depression the larger were these hard grinding stones that, whirled in the grip of the flood torrents, abraded deeper into the rock below. One huge boulder rose out of the river bed with three potholes like steps ascending its downstream side. Water striders skated on the surface film of the largest basin.

Several times we climbed the bank and stood in little forest openings. Once a snowshoe rabbit bolted away. Another time a raven, flying overhead, croaked and veered aside. Many of the openings were strewn with spikenard and meadow rue and star flowers in bloom. Witch hobble lifted its paired foliage like opposing wings and the yellow stars of the Clintonia nodded above the shining green of the leaves outspread below. Small and sweet, wild strawberries had ripened in the sun. We gathered a handful and followed the stream once more, eating the wild fruit as we went and nibbling on the terminal twigs of the black birch, rich with the flavor of wintergreen.

The largest opening we encountered along the upper stream spread away for sixty acres or more. Farmed long ago, now abandoned for half a century, with the forest creeping back across the land, it spoke of pioneer homesteads that once extended far up this forest river. In earlier days, this whole valley was famous bear country. One Maine trapper, who died only a decade or so ago, caught fifteen black bears along the Sunday River.

Now the most arresting feature of the opening was an apple orchard gone wild. It ran for twenty or thirty acres across the clearing. Some of the trees were seedlings, some had sprouted from old stumps, some might well have been a century old. Their trunks were richly robed in lichens. We stopped, at one point, within ten feet of a yellow-bellied sapsucker hammering away in perfect silence. Its strikes, cushioned by the soft, muffling layers of lichen, fell as though on sponge rubber. But sap was still running in the old tree and the sapsucker was there to tap it.

We were sitting at the edge of this wild orchard—making

its last stand with the encompassing forest pressing closer year by year—listening to the whitethroats of the apple trees, some singing in one key, others in another, a few going down the scale instead of up as most do, when we noticed the deer. It was gazing at us from behind a clump of Canada plums, odd-appearing bushes with elongated fruit already forming amid the spiny branches. For minutes we sat perfectly still. Then the doe, relatively unafraid, immensely curious, began stalking in a slow half-circle around us.

It stretched out its neck, sniffed the air, thrust forward its great ears, lifted high its dainty hoofs as it advanced step by step, flipping up in a flash of white from time to time its fluffy tail. At times it was hardly more than thirty feet away. First with one forefoot, then with the other, it pawed the ground, then bent and picked up some special delicacy. Wandering about, browsing here and there, it lifted its head several times and looked around holding a bouquet of red clover blossoms in its mouth. Its slow movements, viewed closely, appeared somewhat stiff and mechanical. It is speed that brings out the grace of the deer family. Now in summer, with the blood-lust of the hunting season lulled by law, the wild creatures grew less wild. At last, still unalarmed and unafraid, the doe wandered away among the apple trees.

Night came slowly on this next to the longest day of the year. Going back upstream in the sunset and the twilight, we tarried along the way. Beside the reach of a quiet pool between two waterfalls, we rubbed on insect repellant and sat for a time watching the leap of trout and the ripple-rings that ran and met and overlapped on the tinted surface of the still water. Minute by minute in the evening air the parade of the insects increased. Frail, flying creatures, born of the rushing water, were on the wing.

Endlessly turning, endlessly followed in every curve and convolution by their reflections, mothlike caddis flies drifted low above the stream. Submerged throughout their early life,

SUNDAY RIVER foams over its rocky bed. With rapids and falls and pools it descends through the Maine forest.

ROBIN at its nest in a northern evergreen. In their last stages as nestlings, the young are nearly fledged.

the aquatic larvae of these insects inhabit little houses of their own devising, tubes thatched with bits of debris or minute stones or fragments of twigs. Some even spin tiny nets among the rocks of the rapids within which they catch their microscopic forms of food. Considering the bizarre and adventurous days of these small creatures as we sat there in the twilight, we reflected on how odd it is that human beings who can hardly wait to discover life on Mars have so little curiosity about life on Earth.

Whitethroats along the riverside now were giving their single-note, crepuscular calls, a sharp chinking sound like the striking of little chisels in the dusk. Slowly darkness enveloped the forest. It settled over the stream. The light-colored boulders and the foam of the river—these were the last to disappear.

We drove back down the rutted road in the night, hemmed in by the blackness of the forest. Frogs hopped away and deer crossed the road before our headlights. And all along the way, unseen in the shadows, veeries sang. We counted more than fifty of these thrushes along the way. A few were uttering their short call-notes, a whirling "Whew!" or strummed "Zurrr!", suggesting the plucked string of a musical instrument. It is this call that Henry Thoreau refers to in his *Journal* as the "yorick" of the veery. We could detect little of the rounded "o" in the sound. It seems more nearly a quick and twanging "Yerrick!"

But most of the birds that night were repeating, without tiring, the refrain of their liquid, slurring song. It came to us from the darkness like a melodious, run-together repetition of the words "deer's-ears-ears-ears." Mile after mile we rode with the veery music. We heard the last of these unseen singers close to the place where we parted company with the stream we had followed with so much interest that day. Only a few hundred yards beyond, it joined the Androscoggin.

THREE

SMUGGLER'S NOTCH

ONCE our summer trip had been a year away, then a month away, then a day away. The tide of time, which nothing alters, nothing halts, nothing hastens, had brought June and the summer solstice and the beginning of our journey. Now we were turning west. We were leaving behind forested Maine and the White Mountains, with their air of other times, of old gentility, about them. We were crossing northern Vermont, with bobolinks jingling up into the sunshine from lush meadows strewn with buttercups and wild blue flags.

Everywhere around us, invisible in grass and herb and bush and tree, flowed the sap of early summer. Tens of thousands of miles of tiny passageways carried it through the surrounding leaves and stems. In imagination we tried to picture how great a lake, how large a flowing stream it all would make.

That night we slept 125 miles to the west of Sunday River. We were among the Green Mountains, close to Smuggler's Notch.

This narrow cleft between Mt. Mansfield and the Sterling Mountains lifts its gray, lichen-covered walls in places to a height of 1,000 feet. We followed their ascent with our eyes, next morning, as they soared almost perpendicularly up and up and up, out of the dusk of the chasm floor into the sunshine far above us. It had rained during the night and the air was sweet and moist, filled with the wild perfume of damp forest mold.

All down the ravine, the broad leaves of the witch-hobble, the small, oval leaves of the water beech, the shamrock-shaped leaves of the oxalis, glistened faintly, each in a sheath of luminous moisture. Hoary boulders lifted the wet velvet of their green moss above shining masses of ferns.

Each section of the country stimulates some special kind of interest. The bare, dry Southwest lies outspread like the pages of a geology textbook. The clothed green hills of New England stretch away, an inviting guide to botany. Here at Smuggler's Notch it is the ferns that attract our particular attention. They were everywhere—at the base of the gray precipices, under the trees, overhanging the edges of the springs, massed around the fallen rocks, bordering the little stream where water thrushes sang, following cracks on the sheer cliff-sides. Vermont is fernland. More than eighty different species of these graceful plants are native to the state. In the pastures and woodlands of two adjoining farms, a botanist once found thirty different kinds of ferns. Appropriately, there is a Fernville not too far from the capital of the state.

Since early days, the vicinity of Smuggler's Notch and Mt. Mansfield has been a happy hunting ground for the student of ferns. It was here, on June 15, 1876, that Cyrus G. Pringle added the beautiful green spleenwort, *Asplenium viride*, to the flora of the United States. For a long time it was thought that this rare fern grew only in Vermont. Later it was reported from a few other scattered stations in northern New England. Always it has been found rooted in the same habitat, among cool, moist and shady ledges of limestone. Its discoverer, Cyrus Pringle, was one of America's pioneer collectors. He added sixteen new species to the lists and, in his latter years, he could write: "My hands have gathered all but thirty-six of the one hundred and sixty-five species of North American ferns."

With their curving, graceful fronds and their infinite number of minute reproductive spores—a single wood fern may produce 50,000,000 spores in a season—the ferns form the dividing line between the lower and the higher plants. Here in the moist and shaded gorge they found a setting to their liking. Each species of fern appears to have its own tolerance for sunshine. Those that thrive in strong sunlight, like the brakes of the open woodland and the hay-scented ferns that cluster around boulders on the pasture hillsides, are in the minority. Most are lovers of the shade. On this June day, although the sun was bright in an unclouded sky, its rays reached the floor of the narrow ravine for only a short time. Long after it was mid-morning in the sky, it was still early dawn in the chasm depths.

As the day advanced we wandered along the notch among the beech ferns and the shield ferns, the polypodies and spleenworts. Their greens were infinitely varied. Their forms ranged from the sturdy to fronds so filmy they seemed made of gauze. In spaces between the green fountain sprays of the fronds, among the clumps and masses and clusters of the ferns, wildflowers grew, red trilliums and violets and the little puffs of white froth of the foam flower, *Tiarella cordifolia*. Under the overhang of one huge boulder, we found blue and white and yellow violets blooming close together.

It was afternoon when Nellie made the great discovery of the day. The name, Smuggler's Notch, dates back to the stormy period that preceded the War of 1812. Contraband goods from Canada passed through this remote defile on its way to the markets of Boston. At one point the base of the cliff is pierced by a small, dank cave where, reputedly, the smugglers stored their cargoes. We were working our way toward this cavern beside ledges, moist and mossy, the home of delicate maidenhair and long beech ferns. Moisture gleamed on the rocks around them. Above us, in a tangle, the voice of a winter wren rambled on and on. It was then that Nellie noticed a fringing of small fern leaves following a crevice in the limestone.

We bent closer to examine them. They were daintily formed and pale green in color. We felt the stalk. It was thin and flexible. We noted its color. It was smooth and green for most of its length, scaled and brown near its base. Point by point we checked off its identification. There could be no doubt. So long after Pringle had first discovered it—in days, as he put it, "when the feelings were young and the world was new"—we were seeing the rare and beautiful green spleenwort.

There was a double pleasure in this discovery. It was an adventure to come upon this rare fern in the very region where it had been added to the lists of science. It was also a delight to find it still rooted here after all the decades of fern collecting the vicinity has known. We left the green spleenwort as we had found it, for others to discover and enjoy. That night we dined in celebration on a dish new to our experience, minted peas. The next morning found us among ferns again, winding upward between dense walls of interrupted and hay-scented ferns, following the twisting, steeply climbing dirt road that lifted us 2,000 feet and more above the green spleenwort of Smuggler's Notch to the summit of Mt. Mansfield.

This loftiest of the Green Mountains, the highest point in Vermont, rises 4,393 feet above sea level. Along the far horizon, on clear days, you can see the Adirondacks beyond Lake Champlain, even Mt. Royal overlooking the city of Montreal. On this morning the air was filled with shining haze. The radius of our view was shortened.

Climbing this New England peak, on this early day in summer, we moved backward in the season, as we had done among the Great Smoky Mountains when we came north with the spring. So far as the yardstick of blooming plants was concerned, our ascent had set back the calendar a full two weeks. It returned us again to the latter days of spring. Along Sunday River, the yellow Clintonia had been in starry bloom. Here it was just in bud. In Smuggler's Notch, the bracts of the bunchberry were waxy white. Here they still retained their leaflike green.

The voice of the wind, that day, was stilled. Hardly a breath of moving air probed the storm-bent firs or swayed the silvery lichen hanging from the oldest branches. We wandered over the wiry mountain grass and the mats of the bearberry and where gray metamorphic rock, at times with a twisted grain like gnarled wood, shouldered up through the thin vegetation. Where sleet and winter gales had broken limbs and shattered treetops, calm now lay all across the face of the mountain. Here the expression, "the face of the mountain," has special significance. For, from many points in the country below, Mt. Mansfield appears to lift in silhouette against the sky a reclining profile complete with forehead and nose and lips and chin.

We sat, that noon, on a shelving ledge where the flank of the mountain tilts steeply downward toward the west. There, watching the play of cloud-shadows across the hazy landscape below, we ate at leisure a lunch of sandwiches, blueberries and a handful of dates. Around us tiny flakes and patches of rock were sloughing off. Bit by bit, slivers of stone were working free. Cracks were imperceptibly growing. Pitted by chemicals of the lichens, chipped by frost, split apart by the wedges of the ice, expanded and contracted by the heat and cold of the seasons, the great rocks were going. If the action of the ages were speeded up to occupy the length of this one summer day, how this solid mountain would dissolve and disintegrate and settle around us! How these towering cliffs of gneiss and schist would fly apart like chaff at threshing time! Each small crack was a part of a long chain of events that leads from the peak to the plain.

For five of its 255 miles, the Long Trail, Vermont's "Footpath in the Wilderness" that follows the crest of the Green Mountains from the Massachusetts line to the Canadian border, runs along the profile of Mt. Mansfield. We followed sections of it north and south. It plunged us into low forests of twisted, storm-wracked trees and carried us around boggy depressions speckled densely with the white tufts of the cottongrass, the sedge that, in the Arctic, provides Eskimos with wicks for their lamps. It led us by humus-rich banks where modest white flowers rose on slender stems above the dark green of three-lobed leaves, flowers that are fertilized almost entirely by a tiny fungus gnat. We pulled away a little of the spongy loam and revealed the roots. Against the black mold, each shone with metallic brilliance. It seemed plated with gold. Appropriately, the common name of this wildflower of the moist woodlands is goldthread.

But the flower of flowers on the mountaintop that day was the painted trillium. All through the mossy forest, under the trees and around the little glades, these great white-and-crimson blooms were scattered singly and in clusters. The three petals, waxy white, stood apart, recurved and wavy-edged. Down the center of each ran an inverted V of rich red that widened toward the base. I sat down on a fallen tree at the edge of a small opening, close beside two perfect flowers. Camera ready, I waited for the sun to emerge from a cloud overhead. Time passed and Nellie wandered away on the trail of a siren bird-sound, an unfamiliar kind of "chewy buzz" that proved to be the call of the junco on its nesting ground.

Minute after minute went by. Overhead a crown or cap of clouds floated just above the mountaintop. The cloud-maker was the mountain itself. Warmer air, rising along the slopes, reached the top, cooled and condensed to form a gray-white umbrella of vapor that remained outspread most of the day. Even when the rest of the sky became almost entirely clear, this floating cap clung stubbornly in place. All around the edges I could see the vapor constantly dissolving and re-forming. Only when the sun sank below the edge of this cloud umbrella and its rays reached the deeply shaded trillium flowers, could I take the picture I desired. So I waited.

Half an hour went by. Nellie came back and started off along the trail in the opposite direction. The sun almost came out, then disappeared, then almost came out again. Each time its pale yellow disk seemed to be breaking through the thinning vapor near the edge, I raised my camera. But invariably new clouds formed swiftly, spreading outward in a swirling mass and dimming the illumination. Thus I waited for an hour, two hours, before the sun finally dropped below the crown of clouds and shone directly on the trillium.

But those two hours were among the most pleasant, the least wasted of the day. Sitting on the trunk of the uprooted tree or wandering about the little glade, surrounded by the silence of that windless June day, breathing in the pure, balsam-scented air of the mountaintop, I was on the edge of wildness and little things of interest were all around me. I investigated the minute world of a knothole with its tiny inhabitants and a microclimate of its own. I laid my ear close to the moss and moldering leaves, listening for little sounds, and once caught the Lilliputian scratchings of a beetle pulling itself over the spongy loam. I explored among the lichen-clad balsam firs, with their wind-tempered branches as unyielding as iron. Everywhere across the forest floor the pageant of decay advanced in silence, leaves moldering, soil being born, the great wheel of change endlessly turning.

As I sat motionless on the tree trunk, a large porcupine came ambling down the Long Trail. It was hardly a dozen feet away when it caught sight of me. It stared undecidedly, then turned and lumbered off up the path, glancing over its shoulder from time to time, its quills erected over its back like an Indian headdress made of spears. Twenty or thirty minutes later the same porcupine came down the same slope of the trail again. This time, when it saw me it made a wide detour through the forest. It seemed put out, in bad humor. Although I heard no sound from it, I could imagine it grumbling to itself as it went.

The sound I did hear, the dominant sound of the little glade early that afternoon, came as a surprise. The voices of the singing birds that intermittently reached my ears I had expected. What I was not prepared for was the rising and falling hum of innumerable flies. No one can appreciate the number of these insects, especially the flesh flies, that inhabit the forests of a mountainside until he sits on a still summer day at the edge of such a glade as this.

All across the opening I saw them resting on the mold, walking over fallen twigs, investigating debris, continually droning through the air on short flights. Most were dark-colored. They ranged in size from species hardly larger than gnats to large bluebottles and gray-black flesh flies. A few hovered in the air over one spot. One of the larger insects walked about on my camera case, perhaps finding remnants of animal fat left there by my perspiring hands. When it took off, passing close to my ear, its buzzing seemed a tremendous sound. It is, no doubt, a measure of the stillness of that day that, for a time, the noisiest creatures on the mountaintop were flies.

Sitting relaxed, aware of all the little sounds around me, enjoying the peaceful calm of these mountain heights, I remembered the young barber who had cut my hair in a small town a few days before. His great ambition, he said, was to work in New York. There was a city! For a good many years, he explained, each summer he had visited his grandfather on his farm in the country. But he couldn't stand it any more. Everything was so quiet! It gave him the creeps. He felt like going out and blowing a trumpet or pounding a drum—anything to make a racket. He represented that new breed, growing in numbers, the Noise Needers.

From the outboard motor to the jet airplane, through the radio and TV, the electric razor and the power lawnmower, almost every mechanical advance has added to the noise of the world. Each successive generation lives in a less quiet environment. In consequence, evolution is at work in massed urban centers. For evolution concerns the present as well as the past, ourselves as well as the dinosaurs. Noise is evolving not only the endurers of noise but the needers of noise.

Those whose nervous systems are disturbed by uproar are handicapped under such conditions. They are less fitted to maintain good health, to endure and to increase their kind than are those who thrive on clamor. What is strain and distraction to one is a stimulant and a tonic to the other. In step with noisier times, the number of Noise Needers is growing. I was told recently of the art editor of a chain of magazines who carries a pocket radio with him all day long and even places it, turned on, under his pillow when he goes to bed at night. Noise is comforting and reassuring to him. He seems in his proper environment when quiet is eliminated. The metallic clangor of rock-and-roll music is, perhaps, symptomatic of the steady rise in the number of Noise Needers. For them, quiet is somehow unnatural, stillness is somehow unfriendly. They feel better, more at home, when they are surrounded by a din—any kind of din. They do not merely tolerate noise. They like noise. They need noise.

Their world, and those who inhabit it, were far away as I sat in the hush of the mountain glade. They were equally remote that evening when Nellie and I walked slowly along the summit paths surrounded by the famous thrush chorus of Mt. Mansfield. The light of the day was retreating. The cloud over the summit was gone at last. Although we were a little late in the month for the main period of song, down from the clinging, cliff-side trees, up from the mountain slopes came the wild song of Bicknell's gray-cheeked thrush. On Mt. Mansfield, this bird nests only above 3,000 feet, close to the timberline of the summit. Its song resembles that of the veery but it is a little shorter, more even and higher pitched. Writing in Arthur Cleveland Bent's *Life Histories of North American Thrushes, Kinglets and Their Allies,* George John Wallace describes it as having "a wild, ringing, ethereal quality that is in perfect keeping with the evergreen solitudes it inhabits."

We sat in the early twilight on the high piazza of the summit hotel listening, as so many previous visitors had done during the Junes of other years, to this evening chorus of the thrushes. Here, too, were the voices of the white-throated sparrows—one small singer more beautiful than the rest. It lifted its voice among the firs below a high cliff. Its tones were more round and rich, its enunciation clearer, the quality of its voice more moving. We remember its strains still as the perfect whitethroat song.

The glow around the mountaintop had faded when we wound downward into the deeper dusk of the lower forests, the realm of the veery and the hermit thrush—Vermont's state bird and, to me, the most ethereal singer of all the north-country thrushes. It was, I think, this latter bird that mystified and delighted Timothy Dwight, onetime president of Yale. In 1821, when Dwight published the story of his extensive horse-and-buggy excursions in *Travels in New England and New York*, he wrote:

"The Meadow-Lark, and particularly the Robin Red-Breast, sing delightfully. There is, however, a bird incomparably superior to either, and to all other birds in this country, in the sweetness and richness of its notes. I am unable to describe it minutely; having never been sufficiently successful in my attempts to approach it, to become thoroughly acquainted with its form and coloring; although I have seen it often. It is a small brown bird; scarcely so large as a robin. Its notes are very numerous; and appear to be varied at pleasure. Its voice is finer than any other instrument, except the Aeolean Harp. What is remarkable in this bird, and I believe singular, is that it sings in a kind of concert, sometimes with one, and sometimes with two of its companions. I have named this bird The Songster of the Woods."

The next day we left the ferns and thrushes of Smuggler's Notch and Mt. Mansfield and all of New England behind. At the top of Lake Champlain we turned west, following an old Iroquois Indian war path along the northern rim of New York State, beside the St. Lawrence River and, eventually, the shore of Lake Ontario. The mood of the world around us, in these green days, had been described seven centuries before by an unknown author in the first words of the earliest song in the English language: "Sumer is icumen in." Summer was "icumen in" all across this beautiful rolling land—coming in with green and rippling grain, coming in with woodchucks sunning themselves on gray whalebacks of rock breaking the sod of pasture hillsides, coming in with new birds of the year trying their wings. In mid-afternoon on the third day, we came to Buffalo and the Rainbow Bridge and the roar of Niagara Falls, the Indians' "Thunderer of Waters."

FOUR

NATURAL HISTORY OF NIAGARA FALLS

TWENTY beams of colored light, 4,200,000,000 candlepower strong, cut through the June darkness. They began at the largest arc lamps of their kind in the world, each weighing more than a ton and rising far higher than our heads on the stone balcony where we stood. They ended, some a third of a mile away, some more than half a mile away, splashing their colors in a sequence of fifteen different hues across the vast curtain of Niagara's falling water. It was ten o'clock. We were on the Canadian side of the falls. The nightly color show, of which visitors the year around never seem to tire, was beginning.

We wandered among the huge, drum-shaped lamps. The hum of electricity, the whir of cooling fans were in our ears. We stopped at frequent intervals to lean against the balustrade and watch the play of colors across the foam of the distant cataract. As we remember that night, the tinted walls of plunging water formed a chromatic backdrop for small and unconsidered events that caught and held our attention. There were gulls—their bodies tinted, too—circling over the gorge where night was being electrically turned into day, moths streaming out of the night toward this battery of great suns ranged along our balcony, tannish caddis flies parading up the front of the lamps, their wings changing hue as each tinted filter slipped automatically into place, sweeping the wave of its new color downward across the lens. They all were part of the life of the cataract, of the ever-fascinating natural history of Niagara Falls.

The day before we had crossed the Rainbow Bridge and accompanied the parade of high-tension lines westward on the Niagara Peninsula of Ontario to Fonthill and Lookout Point and the home of our naturalist friends, the James A. Selbys. A phoebe had built its nest outside the long building of their private museum and a tame blue jay alighted on my shoulder and nibbled at my ear as a sign it wanted to be fed. For years, Al and Gertrude Selby had engaged in the delicate business of banding the minute nestlings of ruby-throated hummingbirds.

Together, late that afternoon, we all drove back to the falls. The object of our attention was a remarkable colony of common terns. They were living dangerously, nesting close to the brink of the Horseshoe Falls. A tapered mass of rock, shaped like a destroyer, cut the rush of the water and extended to hardly more than forty or fifty feet from the lip of the cataract. Here more than 600 terns were nesting. On slender wings, they swirled in white clouds above the rocky islet. Nearby, they plummeted down into the millrace of the waters to emerge with small, silvery fish gripped crosswise in their bills. Once, when a whole school of minnows was being carried over the brink, fifty or more terns milled about close to the edge of the falls. They darted down in arrowy descents and swiftly lifted free of the surface again, often only a few feet back of the spot where the river curved out and down in its thundering plunge with foam and mist and green water intermingled.

On this island stronghold the birds of the colony are safe from predators. Neither man nor beast can reach them. Only winged creatures have access to their rock. From birth, the young terns hatching there are surrounded by the continuous roaring of the falling water. They try their wings in first flights over the rapids and the cataract and the chasm that yawns below it. Their first fishing is done in the seething water where the river, torn among rocks, is making its last rush toward the falls. In this hazardous environment, in a life spent so close to the brink of Niagara and, it seems to us, so close to the brink of disaster, the light and graceful terns were well-equipped to survive.

Long after nine o'clock that night, we heard the clamor of the tern colony above the roaring of the falls. This was in part owing to the shrill, harsh voices of the birds and in part to the low-pitched sound of the falling water. Tests have shown that we never hear the main tumult of Niagara. It is pitched four octaves lower than the lowest chords of the piano. To our ears it is a vibration rather than a sound. And so, too, no doubt, to the ears of the terns, born amid the endless thunder of the cataract, much of the sound around them is inaudible.

We returned to see the colony again the next day, this time with Roy W. Sheppard, retired Canadian entomologist and the man who has studied the natural history of the Niagara region more thoroughly than anyone else I know. Other birds were also active now along the brink. Martins and rough-winged swallows coursed back and forth, often only inches above the tumbling water. At this time of year aquatic insects were emerging from the stream and hastening into the air, sometimes, it seemed, escaping from the water even as it was turning in space and curving downward in its fall. Always along the face of the falling water, mist billowed upward, an airy waterfall flowing in reverse. During nesting time, the swallows often dart out of the sunshine into the mist and emerge with their bills crammed with gauzy insects.

When the big floodlights were originally installed at Niagara, Canadian officials were concerned over the possibility that they might attract unwanted and injurious insects from the American side of the river. Sheppard was assigned to make an intensive study of the species that appeared at the lamps. While he encountered no new economic pests, he found the range of insects attracted was extensive. Most numerous were the moths and nocturnal beetles. The largest insect of all was the giant silk moth, the *polyphemus*. But there were a host of others: flying ants, lacewing flies, snowy tree crickets, roadside grasshoppers, giant water bugs, fish flies, fireflies, cicadas, ladybird beetles. Even that creature of the day rather than of the night, a butterfly—a red admiral, V*anessa atalanta*—appeared at the lights. Incidentally, the phrase, "Broadway butterfly"—like that other descriptive phrase, "brown as a berry"—demonstrates the triumph of alliteration over fact. It is the moth and not the butterfly that comes to the bright lights. Butterflies are usually fast asleep by the time the lights go on.

Because the powerful beams that attract the insects are directed low over the river, they do not lead birds to their death as airport ceilometer lights, pointing straight up, sometimes do. A few years ago, a duck flew headlong into one of the lamps. But this is almost the only time they have been struck by birds. A flock of Canada geese, flying in fog, once veered away just in time and, on another occasion, a second flock landed on the illuminated street below when the birds were caught in sudden fog and rain. The gulls and terns we had seen circling over the gorge in the artificial illumination had seemed neither confused nor blinded by the glare.

During the best part of that long June day, Sheppard showed us places where he had observed special occurrences of natural-history interest in the vicinity of the falls. Here was a sunny recess deep in the gorge where mourning-cloak butterflies gathered in the spring after their winter hibernation. There, herring gulls nested on the rubble of a rock slide. Here was the place where Sheppard had watched eiders diving for crayfish and, up the river there, he had come upon the first otter seen at Niagara in recent times, and here, not far from the *Maid-of-the-Mist* docks, he had observed red and northern phalaropes whirling around and around in their surface feeding. From the floral clock in the park at the Lake Ontario end of the Niagara River to the islands above the falls, our courteous and observant companion pointed out various aspects of nature along the way.

On Navy Island, two or three miles above the cataract, he showed us a bald eagle's nest, now deserted. It was the last nest of the national bird built near the falls where once great numbers of eagles gathered. Alexander Wilson writes, in his *American Ornithology*, of these birds feeding on fish and squirrels and even bears and deer that had been carried over the cataract. When De Witt Clinton published his *Letters on the Natural History and Internal Resources of the State of New York*, in 1822, he reported seeing the greatest concentration of bald eagles he had ever encountered feeding in the gorge at the foot of Niagara Falls. In recent years the only eagles seen in the vicinity have been transients. Great horned owls raised a brood in the abandoned nest on Navy Island one year, and for a time, later on, it was occupied by raccoons.

When eagles were year-round residents at Niagara, they were among the birds Sheppard used to see riding ice cakes down the rapids and toward the falls. Ducks, Canada geese, great black-backed gulls, as well as eagles, would alight on pieces of drifting ice in the calm water above the rapids and then come pitching and tilting down through the white water, navigating a natural shoot-the-chutes, giving the impression of enjoying a thrill ride at a carnival. Each time, before the floe attained the brink of the falls, its passenger would lift into the air and wing its way back upstream again.

Birds that are killed by being carried over the falls appear mainly to be waterfowl that sleep in the river and drift too close to the brink before they awaken. Trying to flee upstream against the swift run of the current, they are unable to reach sufficient speed to become airborne. The most spectacular disaster of this kind overtook whistling swans migrating in the spring of 1912. Between March 18 and April 6, nearly 200 of the great white birds were swept over the falls and perished. Each fall and spring, immense numbers of waterfowl use the gorge below the falls and the Niagara River as a water highway between Lake Erie and Lake Ontario. Since earliest times, the mortality of wildlife at the falls during these seasons has attracted attention. Peter Kalm, the Swedish botanist and pioneer traveler in America, wrote to John Bartram in 1750: "In the months of September and October such abundant quantities of dead waterfowl are found every morning below the Fall, on the shore, that the garrison of the fort for a long time live chiefly upon them. In October or thereabouts such plenty of feathers are to be found here below the Fall that a man in a day's time can gather enough of them for several beds."

When I talked to A. R. Muma, Chief Game Protector for the Niagara region in Canada, he expressed the opinion that most of the birds found dead below the falls have taken off in the gorge, have become blinded and confused in the driving rain and mist near the cataract and have flown into the plunging water. When we donned slickers to ride the *Maid-of-the-Mist* close to the foot of the Horseshoe Falls, we entered a zone where the rain never stops. In this outer fringe of the falls we were surrounded by such gales of wind, such slashing rain, such a thundering tumult of sound that we lost all sense of direction. The confusion of a flying bird is easy to understand. Clifford Keech, captain of the boat, pointed out a ring-billed gull that floated low in the water, riding listlessly with the swirls of the current in the gorge below the falls. We saw several such "waterlogged gulls." They had flown too close to the foot of the falls and had been carried down and tumbled about in the seething water. When they can reach shore, dry out and rest, they recover.

In another zone of rain that never stops, amid great rocks clad in pelts of the finest fur—the silky green alga, *Cladophora glomerata*—near the foot of the American Falls on the Cave of the Winds trip, we observed the birds of Niagara from a different viewpoint. We looked up, as we emerged from the plunging rain, and saw a duck skim low over the crest of the cataract and, high overhead, come scaling down into the gorge. Three grackles followed, lighter birds in the updrafts, seeming to have difficulty with their long tails as they descended. Out on the river, all three apparently alighted on masses of floating foam. A closer glance, however, revealed that in each case the froth had collected around a small piece of driftwood. After the grackles, the next bird we saw was a slate-blue pigeon. With half-closed wings, it slanted along the face of the falls, tossing in the turbulent air, disappearing and reappearing among the billows of mist.

Whenever, in our travels, we have come to a large waterfall we have always found pigeons, gone wild, living around it. Crannies in the rock provide attractive nesting sites. Aptly the original stock of the domestic pigeon was named the rock dove. Here at Niagara, the feral birds nest on little ledges of the sheer walls, far above the spume-streaked waters of the gorge.

To me, the dual character of the domestic pigeon has been a source of frequent surprise. Seen in a flock, swinging this way, turning that way, moving together in precision like a chorus line on a stage, they appear entirely regimented. Observed as individuals, they reveal unexpected independence. They exhibit idiosyncrasies. They are always doing the odd and unanticipated. I remember one pigeon in New York City that, each evening, flew up and up to spend the night at the top of the Empire State Building. Another, for several weeks, came into the tap room of the Yale Club each day to feed on proffered peanuts. A pigeon in the Wall Street district built a window-ledge nest made entirely of rubber bands and paper clips. And at the Little Church Around the Corner, just off Fifth Avenue, the birds that came to feed on rice thrown at weddings developed the habit of tapping at the windowpanes during ceremonies as though impatient for the rice throwing to begin.

During the Second World War, when a British tank commander in North Africa made friends with a pigeon by feeding it scraps of food, the bird is said to have followed his tank all the way from El Alamein to the Gothic Line in Italy. Twice a New Jersey pigeon named Edna sailed from New York Harbor on board the S.S. *Exford* and returned again after living the life of a vacationist on a Mediterranean cruise. Then there was another individualist bird that lived, some years ago, just outside a North Carolina city. Each day a veneer truck passed by on its regular run to a neighboring town. For some unaccountable reason, this snow-white pigeon always joined it, swooping down and flying for several miles just in front of the windshield.

In *Wind, Sand and Stars*, Antoine de St. Exupery tells of a group of Moorish chieftains from the Sahara Desert who were taken to see an ancient waterfall that has been flowing for a thousand years in the Alps. They gazed on and on. Their guide grew impatient and asked what they were waiting for. They replied: To see it stop. Niagara has been pouring over the dolomite-and-shale rim of a glacial basin for at least 35,000 years without a pause. Its slow advance upstream has left behind it a canyon seven miles long and about 200 feet deep. In the long view of the millenniums, all waterfalls are transitory. Niagara, now only a little more than half as high as it once was, may, in the distant future, degenerate into a series of rapids at the Lake Erie end of the river.

But today, its 200,000 cubic feet of water a second, foaming in a vertical plunge of more than 150 feet, forms one of the most famous natural wonders of the world. A million and a half persons a year come to see it. We were now in June, the traditional month of Niagara honeymoons. Vacationists were on the road. Schools were out. All day long, crowds streamed and swirled beside the falls. And all day long, whenever the sun was shining, somewhere, from some angle, rainbows glowed in the mist rising around the waterfalls. In curves and shining fragments, they took shape before our eyes. We saw them in the depths of the gorge. We saw them arching over the brink of the cataracts. Once, near sunset, two terns, fighting in the air, fluttered like white butterflies beneath the curve of a brilliant rainbow gleaming in the misty spray behind them.

Always, of course, we saw those bands of color with the sun behind us, with the white light of its rays broken up into component colors by the prism effect of innumerable droplets of moisture. But at the time, lost in our delight in the rainbow's simple beauty, we gave little thought to explanations. In the out-of-doors, knowing what things are is important. Knowing how things work is interesting. But there is more to nature than the facts of nature. There are beauty and poetry and awe and wonder. Too soon, the child's delight is left behind and "wonder in happy eyes fades, fades away." To forfeit this, to become deaf and blind to all except factual nature, to become absorbed entirely with identifying and explaining merely, is to lose the better half. It is to become as dry and literal and fact-minded as the listener who interrupted a thrilling story about a ship in a storm being driven onto jagged rocks to inquire:

"Were the rocks sedimentary or igneous?"

It is well to view the world at times—to see such things of beauty as the rainbow, the aurora, the cumulus cloud and the butterfly—as the child or the first man saw them. The celebrated American entomologist, Dr. John H. Comstock, one day was telling a group of his students at Cornell University of an exquisite butterfly he had once seen in the Alps. One of the group spoke up:

"What species was it, Professor?"

"At the time," Dr. Comstock replied, "I was not thinking

of its species. I was thinking only how beautiful it was!"

On the third morning of our stay at Niagara, we bade goodbye to the Selbys and crossed to the American side again. On the bridge that carried us over the gorge we recalled another friend of ours and his curious adventure on it. Returning to Long Island from a summer vacation spent gathering botanical specimens and climbing mountains, George H. Peters had taken a short-cut across Canada from Detroit to Niagara. When he went through customs on the Canadian side, the official said:

"It's all right to take these plants out of Canada. But I don't think they will let you in the United States with them."

"Oh, that's all right," Peters assured him. "They all were collected in the United States."

A few minutes later he was telling the same thing at the American customs. But the official was adamant. The plants were coming from Canada and they could not come in. Thinking he would try another bridge, Peters turned around and drove back to the Canadian side. The man he had talked to there had just left for the day. His replacement told him:

"All I know is you are coming from the American side. You can't bring those plants into Canada!"

Faced with the prospect of endlessly going back and forth on the bridge, Peters took stock of his situation. Some of the plants were specimens he had long wanted to add to his collection. He had no intention of giving them up. Waiting until work traffic across the bridge began to mount, he pulled up once more at the American customs. He explained his predicament in detail. He went over it again and again. Cars piled up in a long line behind him. Horns honked. Finally the harassed official waved him through.

"Go ahead," he said, "and get those plants out of here!"

Most of that last day at Niagara we wandered about the mist-watered length of Goat Island, lying between the two great falls. Here, in May, 1861, Henry Thoreau had measured the circumference of bass and beech trees when, on the final journey of his life, to Minnesota, he had stopped off at Niagara Falls. Here he had watched ducks come floating down the rough water of the river, then take off, fly back and ride the rapids again. So, too, on this day, close to a hundred years later, the waterfowl were engaged in this same age-old sport.

Of all the ducks we saw that day, one—the nonchalant mallard—stands out in retrospect. It forms our last sharp memory of the cataracts. When we first caught sight of it, it was being swept by the swift slide of green water toward the brink of the American Falls. Calmly, it checked itself and clambered onto a small rock, awash and partly green with algae. There, less than a hundred feet from where the whole river disappeared and the thunder of the cataract seemed shaking the very rocks, it sunned itself unperturbed. A thin sheet of water streamed over the rock and over its webbed feet. Thousands of great gleaming bubbles, generated in the rapids above, flowed by it, turning and catching the sun.

After preening itself for a time, it slipped into the water again and let the current carry it away. Like a cork it drifted toward the falls, its speed increasing as it went. It showed no nervousness or alarm. It seemed unaware of any peril. We watched fascinated as the distance was reduced to less than sixty feet, less than fifty feet, less than forty feet from the brink of the abyss. The remaining distance seemed to us no more than thirty feet when, with a short, unhurried run, it lifted into the air and went slanting away down into the gorge, a creature of the air finding safety in its wings.

FIVE

MAYFLY ISLAND

A LITTLE after eight o'clock in the morning, on the twenty-sixth of June, a street sweeper in downtown Sandusky was busy shoveling from the gutter into a two-wheeled cart bushels of mayflies. Warm and humid weather had followed wind and rain. From the mud bottom of the shallow western end of Lake Erie the insects, gauzy-winged and trailing threadlike tails, were emerging in numbers beyond counting. A "mayfly storm," one of the early-summer events we had hoped most to see, was building up along the Erie shore.

The shallowest, the muddiest, the warmest and next to the smallest of the five Great Lakes, Erie, over much of its bottom, is a vast incubator of mayfly life. Some years, during the height of the invasion, truckload after truckload of the frail insects is hauled away from Sandusky streets. The arrival of the mayflies varies with the temperature from year to year. The first of the insects appear sometimes as early as May 17, other times as late as June 23. For weeks thereafter the great mating flights continue.

All along the shore of this shallow inland sea, the coming of the mayflies each year brings altered habits to the dwellers in towns and cities. Merchants turn off their neon signs at dusk. Outdoor painting comes almost to a standstill because a freshly coated house soon becomes furry with adhering insects. Traffic slows down on streets that are slippery with the

MAYFLY ISLAND

crushed bodies of the mayflies. On this same day, nearly 500 miles to the west, the bridge over the Mississippi River, at La Crosse, Wisconsin, was closed because of this hazard to travel. Some dwellers along the lake shore disappear for a week or two at the height of the invasion. They are allergic to mayflies. With eyes swollen, red and watering, they are miserable with "Junebug fever" as long as the fluttering hosts remain. In Toledo and other Lake Erie cities, property values are sometimes lower in sections where the insects arrive annually in the greatest numbers.

For the better part of that morning I walked about the streets of Sandusky where merchants were hosing off their windows and sidewalks, where house sparrows were flying up to hover in front of buildings and pick off resting mayflies, where martins hawked for their fluttering quarry amid the traffic, where everybody talked about mayflies as people in other parts of the country converse about the weather.

The street sweeper, a wizened little Italian smoking a short-stemmed black pipe, gave me a version of the origin of the insects that I found was widely held in the region, namely, that they emerge from unfertilized fish eggs. Or, as he put it:

"When fish eggs no hatch, they fly away."

Not long afterwards I encountered a somewhat different version when I fell into conversation with the owner of a store who had just finished brushing mayflies off his awning. His explanation:

"The larvae generate in the residue of fish along the shore."

On a street corner, while waiting for a traffic light to change, an old gentleman with watery eyes confided:

"I tell you, no man knows where they hatch!"

The captain of a lake steamer, who spent his days plowing through the very water from which the mayflies emerged by the million, told me in all seriousness that the insect hosts were born in the marshes of Canada.

"They all come from Canada," he said. "They are blown

clear across Lake Erie by the wind."

When I talked to a waitress in a restaurant she explained that the shooting of big guns at Camp Perry, west of Sandusky, brought the mayflies.

"They always come after heavy gunfire."

Near a bank on a main street, a florid man with a voice of authority proclaimed:

"The origination of the *lava* is a complete mystery!"

The same man pointed to mayflies on the sidewalk and observed:

"Once they land, they never rise again!"

Even as he spoke, mayflies were taking off around his feet without in the least disturbing his conviction of the truth of his assertion.

Somewhat overwhelmed by all this misinformation, I sat for a long time on a green park bench and watched a gray and white kitten playing with mayflies. After a while a man sat down beside me. Cats, he said, rarely eat the insects. There just isn't anything to eat.

"There is nothing to one of those Canadian Soldiers," he declared. "Nothing but wings and tail—absolutely nothing—just nothing."

This name—Canadian Soldiers—is the one most commonly applied to mayflies along the south shore of Lake Erie. Incidentally, along the north shore, in Canada, they are called Yankee Soldiers. Other names we encountered along the way were: lake flies, fish flies, June flies, Junebugs and twenty-four-hour bugs. Along the St. Lawrence they are known as eel flies and on the Mississippi they are referred to as willow bugs. A rather universal idea seemed to be that they are literally the children of a day, that they hatch—with wings—from the eggs, live exactly twenty-four hours and die.

In truth, the clouds of insects that sometimes rise like smoke from the evening waters of Lake Erie have already lived for two years unseen on the mud of the bottom. Only at the very end of their lives are they creatures of the air. Only during the last small fraction of their existence are they visible to us. Before that comes the preliminary life, long, hidden, aquatic, the slow growing up.

Thirty times and more the nymphs molt their skins as they gradually become larger. They spend their days secreted in burrows in the soft mud, coming forth to hunt for small organic matter, feeding most actively at night. Because the great pioneer microscopist, Jan Swammerdam, of Holland, always found muddy particles in the intestines of mayfly nymphs, it was believed for generations that they lived on a diet of mud. Before coming to the surface for their final aquatic molt, they develop dark wing pads. Air, which aids in splitting the shell in this last larval molt, makes the maturing insect buoyant. It rises to the surface, the shell of its skin splits along the back, the winged insect emerges. This event, multiplied a millionfold, takes place in the calm air of the early evening hours.

Three species of *Hexagenia* mayflies take part in the Erie emergences. They are *Hexagenia limbata*, *Hexagenia rigida* and *Hexagenia affiliata*. The first is the most abundant, accounting for more than 12 per cent of the total fauna of the lake bottom. No other insect in the waters of Lake Erie approaches the mayfly in numbers. Always the nymphs on the mud of the bottom are of two sizes. The larger one represents the brood that will next emerge while the smaller, less fully developed ones are the nymphs that will wait until the second year before coming to the surface and attaining wings. Writing in *The Western End of Lake Erie and its Ecology*, Dr. Thomas Huxley Langlois noted in 1954 that of these two groups one is always more numerous. For some reason, perhaps dating from some previous disaster, some crash in population, every second year the mayfly multitudes increase. On even-numbered years, although the same three species emerge each year, their abundance is greater.

That evening, at the end of a long finger of land extending north of Sandusky, we found the Cedar Point amusement park ablaze with light. A living blizzard of mayflies, drifting in over the still water of the open lake, swirled around the amusement rides, the salt-water-taffy stands, the booths where "Presto Pups"—small frankfurters on a stick—were toasting. Each light pole along the boardwalk was furry with mayflies. Concessionnaires were brushing the insects from their stands with turkey-feather dusters. And where a green, glowing neon sign advertised the Breakers Hotel in letters four feet high, thousands of mayflies, tinted by the glow, whirled and spun like a silent display of colored fireworks.

This, however, was only the curtain-raiser for the great mayfly show of the following night. That afternoon we rode a ferry across ten miles of water to Kelleys Island. With its maximum width of seven miles, its shoreline of about eighteen, its more than thirty miles of roads, it is the largest of that small chain of island steppingstones that leads out into the lake above Sandusky Bay. Nearby, in September 1813, Commodore Oliver Hazard Perry won the naval battle that secured the Northwest for the United States at the Treaty of Ghent. Long ago the island's red cedar forests were felled to provide fuel for the *Walk-in-the-Water*, the first steamer to navigate Lake Erie. Centuries before that, about 1625 it is believed, the Indians of the region engraved their history in petroglyphs on what has become world famous as Inscription Rock. And infinitely long before that, granite boulders pressed down by the weight of the ice sheet had carved grooves in its softer limestone, providing geologists with "the finest examples of glacial scouring in the western hemisphere."

Around and around like the hands of a great clock that afternoon and evening we circled the island. Mayflies clung to the tree trunks, bent down the leaves of the roadside plants. Their feet gripped the side of Inscription Rock amid the ancient petroglyphs. They fluttered above the glacial grooves, frail, gauzy, ephemeral and transient, the life of the day above rock that had known the advance and retreat of the Ice Age. Here the great cogs and the little cogs of nature were meeting and overlapping.

Yet here, also, we were face to face with a paradox. Individually knowing a winged life so fleeting that it has become a symbol of the ephemeral, the mayfly, as an insect form, has endured through aeons of change. It fluttered around the dinosaurs and retreated before the glaciers. It is part of that bewildering spectacle of nature, the endurance of the weak. In *La Creation,* the French scientist E. Quinet eloquently expresses our wonderment at this long survival of the frailest forms of life:

"So fragile, so easy to crush, you would readily believe the insect one of the latest beings produced by nature, one of those which has least resisted the action of time; that its type, its genera, its forms, must have been ground to powder a thousand times, annihilated by the revolutions of the globe, and perpetually thrown into the crucible. For where is its defense? Of what value its antennae, its shield, its wings of gauze against the commotions and the tempests which change the surface of the earth? When the mountains themselves are overthrown and the seas uplifted, when the giants of structure, the mighty quadrupeds, change form and habit under the pressure of circumstances, will the insect withstand them? Is it *it* which will display most character in nature? Yes! The universe flings itself against a gnat. Where will it find refuge? In its very diminutiveness, its nothingness."

At the time of our visit the human population of the island was about 600. The mayfly population was astronomical. On these offshore islands the insects sometimes collect along the waterline in windrows three feet deep. At the approach of a visitor, gray ground spiders retreat in waves over this Gargantuan feast. Dwellers on the islands often scoop up the dead insects in bushel baskets and spade them into their gardens as fertilizer. In years of greatest abundance they may see a dark wall of mayflies approaching across the water, carried toward them by the evening breeze. On almost any day during mayfly time the direction of the wind the previous evening will be found recorded in the distribution of the insect hosts on the island. They are concentrated in greatest numbers at the windward edge.

On this particular day, all across the northern portion of Kelleys Island the multitudes of the mayflies were at their high tide. Here, where the shore road left the western edge and turned toward the east, the length of a hayfield slanted down toward the sinking sun. Everywhere across it the stems and leaves of the timothy hay were bending with a burden of clinging insects. Each gauzy wing shone luminescent in the lowered sunbeams. From end to end, the field was swept by captured light, by the yellowish glow of pale, translucent wings.

On each succeeding circuit of the island we stopped here for a longer time to walk for half a mile down the tree- and field-lined road beyond. Cradled on the muddy bottom of the shallow lake, the risen clouds of winged and living creatures had swept ashore and overwhelmed all this countryside. Here we were in the very midst of mayfly multitudes such as we had read about and long had dreamed of seeing.

All the roadsides were furred and fringed with resting insects. All the fence posts, all the telephone poles, all the tree trunks were so densely covered they seemed wearing pelts of mayfly wings. As far as we could see up into the trees, the trunks and boughs and leaves were shaggy with insects. A brown fringing of wings hung beneath each branch, and below the larger leaves clusters of threads dangled down—the filament-tails of the clinging mayflies. Each mass of thorns projecting from the trunks of the honey locusts was festooned with gauzy insects. They covered the dogwood, the poison ivy, the wild raspberries, the Queen Anne's lace, the bindweed, the sweet clover. Willows and rushes seemed lifted from the water, dripping with mayflies. Infinitesimal weight added to infinitesimal weight bent down the leaves of the cattails. Where a green rowboat was anchored near a light, bushels of the brown husks of yesterday's mayflies were heaped on the seats and bottom.

Whenever we touched a bough or bush, a sudden puff of fluttering insects filled the air. Each time I stepped among the roadside weeds to obtain some closer look, I returned covered from head to foot with their clinging forms. Never eating in their adult form, possessing only atrophied mouth parts, these ephemeral creatures neither bite nor sting. What makes them objectionable to the average person is mainly the dead-fish smell of their decaying masses and the oily brown stains, difficult to remove, that their crushed bodies leave on clothing. Before we climbed into the car we always hastily picked off the mayflies that clung to us like burs. They alighted on us almost faster than we could toss them away. The next morning we counted forty-three fluttering about inside the car.

Everywhere, on all the trees and plants, throughout all the unnumbered legions of the ephemera around us, an event was taking place that is unique. Only the mayflies, among all the hundreds of thousands of species of insects, molt again after they attain their winged, adult form. Little winged insects never become big winged insects; once they develop wings their size is fixed. However, when the mayflies lift into the air from the floating canoes of their cast nymphal skins, their bodies and wings are completely sheathed in a thin pellicle. Hours later, sometime before the great mating dance that climaxes their lives, they make a final emergence, leaving delicate molts, frail ghosts of themselves, behind.

Wherever we looked, as evening drew on, this transformation was occurring. In brighter colors, more shining bodies, the imagoes were emerging as we watched. Carefully, through a rent running back from the head, these mayflies, an inch or more in length, pulled themselves free. Last of all, the slender threads of their tails were withdrawn from the sheath of the pellicle. Here and there an impatient one took wing with its shed skin, still attached to the tail filaments, trailing behind it as it flew like a target sleeve towed by a military airplane.

The ash-gray husks of life, the little ghosts this transformation left behind, were anchored thickly on every support. The breeze stirred them and loosened the hold of many. They drifted down in a dry rain around us. They alighted weightlessly on our hands and arms. They reached the road with a sound like the soft sifting of pine needles in a forest. Where a burdock lifted the large green saucer of a leaf, its center was white with accumulated skins. Under every spreading tree the road was coated. The cast pellicles lay like tiny twigs or fruit stems, crisscrossing, piling upon one another over square yards of hardtop.

We picked them up, looked at them through a magnifying glass, marveled at their delicacy and beauty. Each translucent shell was formed of that wonder stuff of nature, chitin. Chitin —pronounced *kite*-in—forms the external skeleton of every insect from the frailest gnat to the armored Hercules and Goliath beetles. It makes possible the shells of crabs in the sea and scorpions in the desert. It possesses remarkable qualities, combines lightness with strength to an amazing degree. When the underwater nymph of the mayfly molts, even the thinnest, almost invisible, chitin sheets covering their gills are shed. Nature has employed chitin more often than any other substance in shaping the small and multitudinous forms of life on earth.

Each delicate jewel of a shed skin that we examined, now empty, had so short a time before been occupied by vibrant life. More and more of the insects, freed of this last encumbrance, rose into the air. They poured up from the grass, fluttered from the tree trunks, took wing from the leaves to join

PAINTED TRILLIUM rising from the moldering floor of the forest high on Mt. Mansfield, in northern Vermont.

FOAM FLOWERS, violets and ferns growing in the moist soil of Smuggler's Notch. Below, the green spleenwort.

the evening dance. An old woodcut came to mind. I had seen it once in a book published long ago. It had caught, as only a woodcut can, the mood and magic of a summer night with mayflies rising into moonlight above a winding stream. Tonight there would be no moonlight. We were in the dark of the moon. But the sunset lingered, the glow of the sky faded slowly, twilight came gradually. And all the while the tempo of the dance quickened—that same primitive aerial dance that had been performed in the dusk of one hundred million summers.

Beside Echo Lake, at Franconia Notch, on the last evening of our travels with the spring, we had paused to watch a little cloud of half a hundred mayflies dancing in the declining sunbeams. Here that half a hundred seemed multiplied by a million. At times the whole island appeared moving and alive. In the wake of the dying breeze the evening was perfectly calm. The insect clouds, mounting from the vegetation, became ever more dense. The mayflies rose and dipped, zoomed and plunged. Up and down, up and down, all in constant motion, the moving bodies shimmered like super heat waves before the landscape.

By seven o'clock the air from the ground to above the treetops was dense with dancing mayflies. Around all the trees we heard the sound of little parchment wings batting against the foliage. When I tried to record the flight in photographs, the insects swarmed into the sunshade of my camera. They alighted on the polished glass of the lens. They made camera work impossible.

Once we saw a red admiral butterfly dash wildly through a cloud of the dancers as though trying to drive them all away. Another time a hen pheasant and half a dozen young birds scaled away into open fields below the main concentration of the mayflies. We followed their flight through our binoculars. Then we became lost in viewing through the fieldglasses the more distant mayflies. Cloud on cloud, each populated by an infinite number of dancers, they extended away across hundreds of acres of open land. Millions—billions—numbers lost their meaning.

More and more we saw mayflies linked together in the air. Usually these paired insects drifted away downward. But occasionally the double forms rose higher and higher among the dancers until they were lost to sight. Into the water of the lake each fertilized female ejects about 1,500 eggs which sink immediately. Usually in less than two weeks, minute nymphs, equipped with tiny gills, hatch from the eggs and burrow into the mud. The long, gradual development of another generation has begun.

In the autumn of 1953, several weeks of reduced oxygen content in the shallow water near the islands killed almost all the nymphs. Around Kelleys Island the mayflies were virtually eliminated. But by the following June nymphs were there again almost in normal numbers. It is believed that under such conditions eggs may delay hatching for several months until the lake water has regained its usual characteristics.

We had circled the island and returned once more to this point of greatest concentration when we witnessed the climax of the dance. Now as we looked up from the roadside, we observed the clouds of insects massed above us like dense billows of black smoke. Whirling in ceaseless motion, their teeming forms blotted out completely the glow of the sky behind them. Our ears were filled with the sound of their multitudinous wings. It swelled, surrounded us, an infinite number of small sibilant noises joining together into a low-pitched roaring that rose and fell like a wind and shower in the forest. We felt exhilarated. Never before had we been in the midst of such a floodtide of vibrant life.

I remember once we stooped and freed a mayfly entangled in the grass. It flew hurriedly away to join the dancers. Its progeny may live, may owe their lives to this act of ours. Why did we do it? We could hardly say. Here life was abundant, life was cheap. One more among so many—what could it matter? Perhaps our reason was that we were on the side of life and in so small a degree we had altered the balance of the world.

It was a little before nine when the height of the dancing came. Half an hour later the insect clouds were thinning away. The gusty roaring of the many wings was reduced to a murmur. Although the dancers would continue to whirl on and on in diminished numbers, the main flight of the night was over. The event that climaxed the long growing-up for an infinite number of mayflies had come and gone. How soon it all was over! Tomorrow would find their lifeless bodies floating on the lake in growing rafts or piled upon the land, drying in the sun. All of them, so soon, so soon, had known and left behind their joyous hour, had met and tasted and embraced and then had lost almost at its beginning what W. H. Hudson once called "the feast and fairy-dance of life."

Slowly we circled the island for a final time. Where the lighted south road leads to the ferry dock, each street lamp was the hub of a wheeling, spinning cloud of mayflies. On the ground beneath, the insects piled up, already heaped into mounds a foot deep. From top to bottom, an illuminated outdoor telephone booth was plastered with clinging mayflies. When we paused momentarily beneath a street lamp the white hood of our car was almost instantly covered.

What purpose is served by the mayfly's passion for the light? All around the lake that night, hundreds of thousands of lamps were holding the insects prisoners. Instinct seemed defeating itself. The pull of light seemed stronger than any other force. The fact that all these lights were man-created may supply the answer. They are only recent additions to the mayfly's world. Some instinctive predilection for the light that would, for instance, draw them higher among the trees at twilight or pull them to the lighted surface from the muddy bottom of the lake, may have played its part in the long history of their kind. Man and his inventions are, after all, comparatively modern innovations. The mayfly had lived for infinite stretches of time before either appeared. As so often is the case, what we are observing is the functioning of an age-old instinct suddenly confronted by man-produced changes, an instinct that was originally developed and adapted to a manless world.

When I glanced at my watch, I saw it was nearly ten o'clock. We hurried toward the dock. It was time for the next-to-the-last ferry to leave for the mainland. But we saw no boat. We saw no lights. Then we caught sight of a dark form moving silently away into the night. We had expected to see the glitter of many ferry lights. But there were none. The boat was departing furtively, like a vessel in wartime, stealing away with darkened decks across the mayfly-infested waters. We sat in the night, wondering if it would return, until after eleven o'clock. Suddenly it loomed out of the darkness. Plowing through water completely mantled by a churning blanket of fallen mayflies, it maneuvered to the dock. The white beam of a spotlight shot along the ramp, lighting our way. A cloud of insects swirled around it. We rolled onto a deck thickly carpeted with mayflies. The captain climbed hastily back to the pilothouse, brushing off the insects as he went, and slammed the door. Diesels throbbed below us and we pulled away. Instantly all lights, except necessary navigation lamps and a single spotlight pointing straight ahead, blinked off. So we began our darkened passage in a mayfly blackout.

The stars were bright in the sky above us. Stepping from the car, I clutched the rail just in time as my feet shot out from under me. The whole deck seemed thickly coated with grease or slippery oil. Hand over hand, as though on a ship in a storm, I made my way forward. Around the single white eye of the spotlight, the superstructure of the ferry was hidden beneath masses of hanging insects. On the deck below there rose a considerable hillock or rounded drift of fallen mayflies.

And all over the forward deck beyond I discovered an extraordinary sight. Thousands of mayflies had landed there. But they rested in no haphazard fashion. In the manner of a partially opened fan, the multitude spread out as the path of the light widened. All the insects pointed inward toward the source of the illumination. Their bodies, each a fragment of a radiating line, suggested iron filings in a magnetic field. They all remained motionless, charmed by the sun of the single spotlight.

Even before we approached the dock, deckhands were already sweeping away the mayflies. The accumulating grease of their crushed bodies becomes progressively more dangerous as the evening advances. Each night, all through the mayfly season, the ferry is thoroughly hosed off at the end of the final run.

One o'clock had come and gone before we reached our mainland cabin. Late as it was, we were an hour or more going to sleep. Our minds still teemed with memories of all those clouds upon clouds of dancing life.

SIX

THE MICELESS HOUSE

DAY after day now we saw no newspapers. We heard no radio. The world wagged on without us. In June how good is the news when you get it only from nature!

There is an old Bohemian legend that tells of twelve silent men—the twelve months of the year—sitting around a fire that never goes out. The fire is the sun. Three of the silent men wear cloaks of green. They are the months of summer. Wherever we rode in these days of the Green Men, green was the cloak of the earth around us. This was the time of grass, the time of leaves, the great chlorophyll-producing time of the year.

A single apple tree may hold out to the sun 100,000 leaves; a single elm more than 1,000,000. A sugar maple may be clothed in half an acre of foliage. Exclusive of Alaska and Hawaii, the United States comprises 3,022,387 square miles. Who would dare guess the area of the summer leaves this land supports? Our imagination comes to a standstill when we attempt to add leaf to leaf, grass blade to grass blade, across the 3,000 miles of the mainland.

Amid the mayflies of Kelleys Island we had reached the 2,000-mile mark of our trip. Now as we ran west along the southern shore of Lake Erie, the sunlit day was filled with the clatter of mowing machines. Alfalfa and red clover, rye grass and bromegrass, red top and timothy lay drying in swaths or ran in windrows or dotted the hayfields with rectangular bales.

THE MICELESS HOUSE 51

Wherever the road led us we breathed air sweet with the perfume of new-mown hay. This nostalgic scent, so symbolic of early-summer days, has been traced by scientists to a chemical called coumarin. Within the bodies of certain microorganisms, when sweet clover hay is harvested moist, coumarin is synthesized into a related compound called dicoumarin. It has the property of reducing coagulation of the blood.

Years ago I asked the eastern botanist Dr. Harold Moldenke a question that no doubt has occurred to many a man as he mowed his lawn. Why does grass continue green and healthy in spite of repeated shearings, while most other plants turn brown and die when cut? The grasses and the sedges, he told me, have the growing point of the stem at the base. In consequence the stem is continually pushed up from the bottom. Mowing or grazing by livestock merely cuts off the tops of the stems and leaves without injuring the growing points below. In fact, growth is often stimulated by removal of the upper part of the grass because it permits the sunlight to penetrate to the growing point with more intensity. When grass is uncut, on the other hand, the shade of the leaves tends to decrease the rate of growth. It is almost entirely leaves alone that we see in a lawn or close-cropped pasture.

The dependence of all animal life on the grass and the green plants of the earth has never been expressed more eloquently than by the Greek philosopher Heraclitus. "The rain falls," he wrote some five centuries before the birth of Christ, "the springs are fed, the streams are filled and flow to the sea, the mist rises from the deep and the clouds are formed, which break again on the mountainsides. The plant captures air, water and salts, and, with the sun's aid, builds them up by vital alchemy into the bread of life, incorporating this into itself. The animal eats the plant and a new incarnation begins. All flesh is grass. The animal becomes part of another animal, and the reincarnation continues."

Part of this incarnation appeared amid the Ohio hayfields in the form of the dispossessed mice that, from time to time, scurried across the roadway. We encountered them at intervals throughout the haying country. The disappearance of the protecting grass had forced them abroad even in the glare of the sunshine. Several times we saw them pass each other as they raced in a panic in opposite directions across the highway. Suddenly their hidden world of paths and tunnels was laid bare. Suddenly the protecting walls of the grass fell about them. Suddenly they were left roofless and exposed.

Circling around and around, starting at the edges and working inward toward the center of the field, a mowing machine lays down swath after swath of fallen grass. As it circles, the mice retreat inward before it. At each turn the concentration in the remaining grass becomes greater. The last stand of hay, at the center of the field, swarms with the small rodents that find there a temporary sanctuary such as the sea birds find in the calm at the eye of a hurricane. Then, before the shuttling knives, the last stand falls. The mice scatter in terror, disorganized and dispossessed entirely. Thus their world is changed on a sunny day in summer.

That world, beneath the canopy of the growing grass or under the winter snow, is centered in a maze of paths. They crisscross endlessly and reach every portion of the field. In a single square mile of rich grassland, a scientist once calculated there were 100 miles of mouse paths. With yellow-tinted cutting teeth, razor-sharp, the little animals—dark brown of fur, beady black of eye, stubby of tail, chunky of body—keep their rodent highways cleared and in good repair. Each straw is removed. Each grass stem is clipped off close to the roots. The result is a smooth runway about the same width as the track of a garden hose that has long lain in the grass. Rarely do the field mice leave the labyrinthian maze of their pathways. At intervals along these public thoroughfares the cleanly little creatures establish communal toilets where they deposit their small green pellets of dung. Although the winding,

THE MICELESS HOUSE 53

bisecting trails lead to every part of a field, each individual mouse keeps pretty well to its own established range, a tenth of an acre or less in extent.

The mice we glimpsed racing for cover among the cut-over hayfields appeared to have hardly any ears at all. Close-set and small, their ears are almost entirely hidden in the fur. Beneath the coarse guard hairs that shed water, the pelt of the meadow mouse is soft and silky. It might well have provided the material for those celebrated mouse-skin breeches that were the delight of Gulliver's Emperor of Lilliput.

Known variously as the field mouse, the meadow mouse, the bear mouse, or—as it is referred to in England and appears on the lists of science—the vole, the scurrying creature we saw belonged to the genus *Microtus*. This genus occurs throughout the temperate zone of the Northern Hemisphere. Nearly 100 different species and races are found in North America. They range from sea level almost to timber line. The mouse of the Ohio fields, a species found from Maine to Minnesota, the largest mouse of the Great Lakes region, is *Microtus pennsylvanicus*.

Sifting the deposits of loess in Nebraska, paleontologists have found the skulls of *Microtus* along with the remains of the sabre-toothed tiger, the giant ground sloth and other extinct inhabitants of the Pleistocene. Through aeons these mice have survived in spite of the unending attack of the longest list of enemies known to any small mammal.

Ranged against them are the foxes, the skunks, the weasels, the badgers, the wildcats, even such huge predators as the Alaskan grizzlies. Aristotle tells of the ancient practice of turning swine into a field to root mice from their tunnels and runways. Hawks and owls drop on mice from the sky. When corn shocks are overturned at husking time, the birds of prey sometimes swoop close beside the working men to snatch up the fleeing mice. In Montana, beside a field where mowers were at work, John H. Storer once saw a hawk waiting on almost every fencepost. Crows often walk back and forth across cut-over hayfields hunting for mice. When they take to water, the little rodents not infrequently fall victim to bass and pickerel. Blacksnakes, milksnakes and other reptiles live largely on mice. There are even records of an insect, a praying mantis, killing a field mouse.

Against this formidable array of enemies the mouse pits its adaptability, its capacity for living in varied habitats, its wide choice of foods and its prodigious fertility. It is at home in pastures, orchards, gardens, grain fields, hayfields, salty grasslands at the edge of the sea, even in vacant lots in great cities. It can live on dry, sandy soil but by preference it chooses moist, even swampy, areas. When it lives near streams it dives readily into the water to escape pursuit. At times meadow mice have been seen swimming leisurely among pickerelweed and at least on one occasion beneath the ice in shallow water. Victor Cahalane, the American mammalogist, tells of seeing one of these small rodents leap into a swift mountain stream and go paddling away around a bend. It is during such aquatic adventures that the animals fall prey to their enemies swimming in the water below.

If it emerges safely on land, the field mouse gives itself a few shakes, sending droplets flying, and its fur is at once fluffy and almost dry. The globular nest of dry vegetation that the rodent weaves in a few hours' time also sheds water well. So interlaced are the stems and blades of the dead grasses that the interior is essentially waterproof. Within, where the young are born, the nests are lined with cattail or milkweed fluff, with the softest material available.

The baby field mouse is born pink, naked, blind, deaf, helpless. It possesses, however, an amazing capacity for enduring long submersion under water. A newborn mouse once revived after being under water for more than 30 minutes. In times of flooding, whole litters sometimes save themselves by clinging to the swimming adults, in this way being transported to

THE MICELESS HOUSE

safety. At birth the weight of a mouse is about that of a penny. But its growth is swift. When it is full grown it may weigh four times as much as a silver dollar. At the end of a week it has teeth and is covered with velvety fur. Two days later, with eyes open, it is nibbling on tender leaves. Before it is two weeks old it is weaned and at the age of three weeks it is on its own, running along the trails that wind through the grass near the place of its birth. If it is a female, it is ready to breed. And only a few hours after giving birth to from four to eight young—the average is six but the number sometimes is as high as thirteen—a female mouse is ready to breed again. The females are promiscuous and the breeding season extends to almost every month of the year.

Some years ago, in a high-school science class at Amboy, Indiana, the teacher, Wendell Lamb, rigged up a ten-inch wheel in a mouse cage. To the axle he attached a "mouseometer" to record the number of turns produced by each mouse's run. In an average day, he found, one of the lively little animals would spin the wheel 15,000 revolutions, the equivalent of a straightaway run of about two and a third miles. One female field mouse was running on the wheel at midnight. She stopped for an hour or two to give birth to eleven baby mice. Before dawn she was back on the wheel, running again.

Compared to the staggering fertility of the hosts of the mayflies we had just seen, these mice of the fields seem far-scattered and few in number. Yet among mammals, large and small, their fertility is unequaled. Without natural checks they would doom themselves. Vast numbers of mice are destroyed by innumerable enemies. But a continual flood of new mice takes their place. As a species, the field mouse survives to a major extent through the rapidity of its reproduction.

Vernon Bailey, the government biologist who studied mice extensively in captivity, recorded one female that, in a single year, gave birth to seventeen consecutive litters. They appeared on May 25, June 14, July 8, July 29, August 23, September 18, October 18, November 9, November 30, December 21, January 12, February 2, February 23, March 18, April 8, April 30 and May 20. A daughter of the first of these litters produced thirteen litters of her own before her first birthday. Bailey calculated that if only four litters a season, each containing the average of six young, followed each other for five years with all natural checks removed—which, of course, they never are—the progeny of one pair of meadow mice would grow to more than 1,000,000 mice. The whole life of this little animal appears speeded up—its heartbeats, its activity, its growth, its feeding, its reproduction.

Beyond Toledo, as we rode north into Michigan that day, the land rose gently, the hills rolled away, the pink trumpets of the bindweed and the broad, flannel-like leaves of the mullein decorated the roadside. Once we passed a meadowlark without a tail hunched on a fencepost like a small balloon of feathers. Another time when we pulled to a stop a killdeer circled above us crying, swerved away, swooped, touched the ground, ran half a dozen steps and called again. In its perfect contact with the earth, in every bird's change from an aerial to a terrestrial medium, we see a hundred million years of trial and error compressed into the final result.

Our advance that afternoon carried us through a land of green pastures, timothy and red clover, orange hawkweed and, beside the road, the silver globes of the poised, winged seeds of the goatsbeard. Here, too, all along the way, haymakers were active under the sun. The smell in the air was the smell of new-mown grass; the sound in the air was the sound of the mowing machine. It was a smell and a sound that must be associated in a million little mouse brains with a time of change and crisis. During the rest of the year the rodents meet only the ancient enemies, the old, old problems of their kind. But in these summer days they see their world laid waste by a new enemy, the machines of man.

THE MICELESS HOUSE

Over the years our knowledge of these diminutive rodents has been built up by experiments in the laboratory and observations in the field. In England meadow mice have been tagged with radioactive material and then followed in their invisible movements by means of Geiger counters. On the other hand, the American authority on field mice, Dr. William J. Hamilton, of Cornell University, made many of his extensive observations simply by sitting quietly in the low crotch of an apple tree early in spring, before new grass hid the runways, and watching the normal activity of the animals below him.

That activity extends throughout the twelve months of the year. The mice appear to get their rest in short naps, for they are found abroad at all hours of the night and day. As a rule, however, they are most in evidence in the evening and during the early morning hours and are least active at midnight and midday. They neither hibernate nor, except in some far-northern regions, store up food. During the winter they often make long excursions beneath the snow to reach corn shocks, haystacks and orchards. In January, J. P. Lindhuska examined 315 corn shocks at the Rose Lake Wildlife Experiment Station, in Clinton County, Michigan. He found they held a population of 479 mice. By spring, other investigations have revealed, the mice may have consumed nearly half the corn left over-winter in shocks.

But it is during the summer months, when food is most abundant, that the wild mice indulge their prodigious appetites to the full. Roots, leaves, berries, seeds, bulbs, bark—even an occasional insect or snail—form their menu. The rodents masticate their food so finely that oftentimes even the cell structure is destroyed. In studying the stomach content of mice, government scientists frequently have trouble determining exactly what has been eaten. The bulbs of wild onions have been detected by smell and the bark of trees by color.

In the course of twenty-four hours, one of the meadow

mice we saw along our way that day will consume its own weight in vegetable matter. During a full year it will eat between twenty-four and thirty-six pounds of food. Not infrequently there are 1,000 mice in a single meadow. It takes twelve tons of grass to sustain them for a year. During the time 100 mice in an alfalfa field are consuming 300 pounds of the clover they are destroying something like 3,000 pounds more in getting the tender upper leaves they relish most. For it is their habit to bite off the stalk close to the ground and then feed only on the most succulent terminal leaves.

Many a farmer working in his hayfields has been puzzled by coming upon little piles of crisscrossed, match-length sections of timothy stems. They are by-products of the field-mouse feasts. Intent on gaining the seedhead at the top, the mouse snips off a stem near its base. The severed grass remains upright, supported on all sides by the thick stand of hay around it. The mouse rears up on its hind legs and nips off the stem higher up. The grass drops a couple of inches. The process is repeated again and again until the seed head has been lowered within reach. At the end of its meal, the mouse leaves behind the little pile of stem sections almost uniform in length.

In the Great Lakes region through which we were traveling the field-mouse population rises and falls in more or less regular cycles. Each three or four years it attains a high point. At such times there may be 200 mice to the acre. Synchronized with this mounting population, mites, fleas and lice also multiply. As many as 1,000 mites cling to a single mouse. Such parasites, together with the predators and, especially, epidemics of disease level down the population once more. Even under normal conditions the life of the mouse extends over one or two years at most. Heavy frost after a day of rain in fall may kill virtually all the mice in an area. Trapping records show that comparatively few adults survive through the winter.

Lived intensely, filled with active days and active nights, the span of the mouse soon reaches its end. But while it lasts, while the continuity of this chain of rodent life is maintained through the swift replacement of individuals, *Microtus* represents one of the most important living threads running through the life of the fields.

The successive mice we saw and the dramatic changes that were overtaking them all along the way—these were much in our conversation as we drove north through Temperance and Azalia, Milan and Ann Arbor to end our day's journey at Fenton. This was a homecoming for Nellie. For here she spent summers in her childhood. And here we had a reunion with her uncle, Clifford J. Phillips. And here, as an odd conclusion to this day with mice in the mowing fields, we came, as we rode about Fenton in the evening, to "the house without a mouse."

This white wooden structure, with twelve rooms and numerous gables, sat well back among shade trees at the corner of Shiawassee and Adelaide Streets. The builder, Andrew J. Phillips, a more distant relative of Nellie's, had grown up on a farm. There he had developed an aversion to field mice in particular and all mice in general. When his window-screen factory prospered and orders came in from all parts of the country, even from the White House in Washington, he built the largest home in town. And he remembered mice. He decided that no rodent should be allowed to enter his house. Each wall was constructed with two-by-fours piled one on top of the other to fill the space completely between the lath and plaster and the outer walls. Nowhere in the structure were there any passageways through which a mouse could scamper. The building, completed in 1893, is now a nursing home. For more than half a century it has remained uninvaded by rodents, known through the region as The Miceless House.

SEVEN

EIGHTY-FIVE MILES TO BREAKFAST

UP and over and down in a 1,000-mile swing, in the days that followed, we circled the edge of the Lower Peninsula of Michigan. All along the way names on the map—names with romantic, far-off pioneer associations, names that echoed in memory like Thunder Bay and Presque Isle and the Straits of Mackinac, like Manitou Island and Manistee and Charlevoix—became transformed into definite images, sharply defined scenes that would remain in our minds.

A glance at the map of Michigan shows that the Lower Peninsula resembles a broad lobster claw or, as is more often pointed out, a wide mitten. Our first day carried us up the shore and around the thumb of the mitten. For 200 miles and more, past Grind Stone City and Hat Point and Wildfowl Bay, we drove through falling rain and clinging mist. In the Great Lakes region, on the average, there are eleven days of rain in June, eleven in July, nine in August and eight in September. It was still drizzling that night when near Bay City, at the end of Saginaw Bay's fifty-mile invasion of the land, we fell asleep with frogs calling from flooded fields around us.

In the Indian tongue from which it is derived, the name Michigan means Lake Country. Two-fifths of the state's surface is water. If the Great Lakes, representing one-third of the world's fresh water area, are the giant puddles left by the glaciers, the smaller bodies of water scattered across the mainland of Michigan are the droplets. This whole area is glacier country. The handiwork of the Ice Age is everywhere. With water never far away, it is cloud country, too. All the sky, on the day after the rain, was sown with wind-blown clouds, dazzlingly white. As we advanced north along the Huron shore we were constantly reminded of the summer fishing. Bait stands carried such signs as: "Nightcrawlers in Moss," "Crickets and Nightcrawlers," "Worms for Sale—Day or Night." And beside the road, in farmyards, children were playing with puppies and kittens, the year's new installment in the succession of life.

Half an hour or so after we stopped for lunch—and ended with that early-summer dessert of the midwest, rhubarb-and-strawberry pie—we came to Rifle River. In pioneer times, this stream flowed through forests, the swiftest river in Michigan. It hurried toward Lake Huron, rushing between bush-lined banks, clear, cool, famed for its abundant trout. Axe and plow, in the years that followed, stripped away the shade, let in the sun, warmed the stream and filled its water with silt. Four-year trout, which averaged sixteen inches in length in the unpolluted Au Sable and Manistee rivers, averaged only twelve in the muddy Rifle.

Then, in the spring of 1950, the Michigan Department of Conservation commenced a grand experiment in rehabilitating a river. Through the use of funds obtained from fishing licenses, it began altering the stream from one end to the other. Here the conservation scientists speeded up the current. There they planted stands of willows along the bank. Here they bypassed a stretch of warm water. There they created pools and spawning places. The whole watershed of the river, embracing approximately 100,000 acres of woods and fields, was included in this pioneer experiment in changing the fish habitat of a Michigan stream.

In two years, the workers stabilized 113 eroded banks, produced eight miles of grass waterways, erected 106 structures to provide better feeding and spawning conditions, planted 400,000 trees along the banks for shade and cover. Farmers in the region altered their methods of agriculture to prevent erosion. Check dams blocked the runoff in gullies. Already, as we stood beside the stream, the silt had been reduced by almost half. As the willow cuttings and pine seedlings and streamside bushes attain their growth, the river will steadily approach its character in a former time. And healthier fish will dart through its clearer water.

Other streams—the Tawas, the Au Sable, the Thunder Bay, the Ocqueoc, the Cheboygan—crossed beneath our highway, each close to the end of its journey. On our right the glitter of Huron waters appeared and disappeared among the trees. Far out, trailing miles of smoke behind them, ore vessels headed up and down the lake, to and from the great funnel of the Soo Locks. Once we turned inland where, in the jack-pine barrens around Mio, Harold Mayfield showed us the first recorded nest of the black-backed or Arctic three-toed woodpecker on the Lower Peninsula of Michigan. We watched the parent birds come swooping in to alight at the nesting stub while all around us the jackpines rang with the song of the rare Kirtland's warbler. Near Alpena we came to the Forty-Fifth Parallel and for an instant we were precisely equidistant from the Equator and the North Pole. Somewhere beyond we pulled up to follow a deer trail through sun-splashed open woodland, among bracken and sweetfern. Then our road ran on and on, leading us northward into a land of longer twilights.

One morning when we were on our way at dawn we hunted a long time before we found a wayfarer's breakfast. Each small community we rode through was sound asleep. No one was abroad. The sidewalks were empty. Doors were closed. No sign of life stirred around any restaurant. We continued from village to village and small town to small town. Not infrequently we stopped to examine some wildflower or to listen to

EIGHTY-FIVE MILES TO BREAKFAST 63

the dawn chorus of the birds or to enjoy the early sunlight on aspen or birch or spruce beside the way. When at last, in that sparsely settled land, we found the doors of a restaurant open, we had driven eighty-five miles before breakfast.

That was the day we reached the top of the mitten, curved around its end and started down the other side. It was also the day when we invented words. Nellie started it with what seemed a perfectly logical observation that if there is a word "herbaceous" there ought to be a word "shrubaceous." This was followed by deciding a good name for a combination swamp and bog would be "swog." And it all ended in hastily changing the subject when I volunteered that if a small lion is called a cub and a small horse is called a colt a small swallow might be called a "sip."

There is, in the far northwestern corner of the Lower Peninsula, at the western gateway of the Straits of Mackinac, a lovely area of wilderness. Michigan has preserved it as the Wilderness State Park. In our minds it is associated with the indescribably delicate perfume of small, bell-like, pink-tinted twinflowers that nodded in pairs among the moldering logs and the mossy glacier boulders of the forest floor. These "twin sisters" of the north-country woods, rising from trailing vines that creep among the shaded mosses, are related to the honeysuckle. They were the special favorites of the great systematic botanist, Carl Linnaeus. His affection for them is recorded in their scientific name, *Linnaea*.

We were north of Petoskey, following the shore road, when we climbed through woodland onto an immense promontory that overlooked the upper end of Lake Michigan. Gazing out over the white-flecked expanse of blue, I felt an old excitement stirring within me. This was the first large body of water I had ever known, the lake of my childhood. Our first river, our first lake, our first mountain, our first forest—none encountered later ever leaves an impression so indelible.

At last we turned and let our eyes wander over the wooded

land that stretched away as far as we could see into the interior. The sky above all this forest land, above our promontory, above the waves of the lake shallows had been cut and crisscrossed, hardly a century ago, by the million passing wings of the now vanished passenger pigeon. Except for being larger, having a blue-gray head and lacking the black spot behind the eye, this bird closely resembled a mourning dove.

The place-names of the region—Pigeon Lake, Pigeon Hill, Pigeon River—are echoes of the days when the uncounted hosts of these wild birds nested here. In pioneer times a single continuous nesting area near Petoskey covered more than 260 square miles. Across Lake Michigan, another nesting area extended for 100 miles through central Wisconsin. Literally obscuring the sun, the migrating flocks streamed north and south in flowing rivers of birds. Alexander Wilson tells of one such living torrent that passed above him hour after hour near the Kentucky River. He calculated it was at least a mile wide and 240 miles long and contained more than 2,000,000,000 birds. No one can even estimate how many billions of passenger pigeons once lived in this land where not one single bird of its kind breathes today.

So densely packed were the roosting birds that small oaks bent to the ground with their weight. They often rested one upon another. Sometimes the masses were a dozen pigeons deep. Cotton Mather describes vividly such a roosting place: "Yea, they satt upon one another like Bees, till a Limb of a Tree would seem almost as big as a House." As many as a dozen birds were reported killed by a single rifle ball fired into such a mass. And the firing continued incessantly.

Meat hunters in one small area in upper Michigan slaughtered 700,000 passenger pigeons in a single year. Salted and packed in barrels, their bodies went to market. Live squabs, tens of thousands of them pulled from the nests and placed in wooden crates, traveled by express to the larger cities. Ruts in the forest roads were sometimes filled in with tons of the discarded pigeon wings. Netters attracted the passing flocks by using as decoys birds with their eyes sewed shut. In the whole revolting story of man's inhumanity to fellow creatures of the earth, the record of the passenger pigeon forms one of the darkest pages.

While the State Senate of Ohio was receiving a report that concluded that the passenger pigeon would never become extinct, and Massachusetts was enacting a law to protect the netters of wild pigeons, the slaughter continued. The railroad and the telegraph hastened the end. Station agents, with their dot-and-dash messages, reported the arrival of the migrants all along the line. The railroads ran special trains for the gunners and the express companies shipped back the birds they killed. The slaughter of the pigeons became big business. Yet so great were their numbers that little decrease was noticed until about 1880. Less than four decades later, however, only a single passenger pigeon remained alive. This was Martha, a caged female at the Cincinnati, Ohio, Zoo, now one of the mounted specimens at the National Museum, in Washington. When she died, on September 1, 1914, a whole species had been destroyed; the teeming, vibrant life that was the passenger pigeon's had disappeared forever from the earth. No remnant remained in all the woods that spread away below us where once the multitudes had fed and nested.

One summer day, years ago when I was a small boy, I lay in the warm sand at the top of a high Indiana dune and let my eyes follow the great curve of the Lake Michigan shore. When it became dim, then lost entirely in the summer haze, I followed it still, traveling on and on to circle, in imagination, the whole long body of water. Now, so many years afterwards, as we rode south around Grand Traverse Bay, out on the Leelanau Peninsula, past the Sleeping Bear Dune, then inland down the center of the state, we had started to draw, in reality, that imagined ring around a lake.

The 22,400 square miles of Lake Michigan's water extends

for 300 miles in a north-and-south direction. No other body of fresh water in the world, with the possible exception of Lake Baikal, in Siberia, stretches so far up and down the map, embraces so many degrees of latitude. Traveling from north to south down its eastern side and from south to north up its western side, we saw the changes wrought by four degrees of latitude on the fauna and flora around us.

Before we reached Big Rapids we had left the dominant aspen country behind. Beech and elms grew more numerous. By the time we were halfway down the lake we had passed beyond the farthest limits of the beaver and the black bear. By the time we were three-quarters of the way we had also left behind the range of the bobcat and the porcupine. Coming into Grand Rapids, we passed a Ten-Mile Road, a Seven-Mile Road, a Three-Mile Road. It was near the latter that we saw the first of the innumerable Michigan vineyards that extended away around us through much of the southern part of the state. By now we were meeting black walnuts and box elders and an occasional sassafras tree with leaves mitten-shaped like the peninsula itself.

Elderberry bushes, as we moved south, provided a flowering yardstick with which we could measure changes in climate and growing season. In the upper sections of the mitten, the bushes were just showing white. Long before we reached the southern border of the state, however, they were in full bloom, covered with the snowy masses of their flower clusters. Among the cornfields the new plants rose higher and higher as we advanced. Strawberry pickers increased and rows of rounded cherry trees became more thickly sprinkled with brilliant red. In the coming of spring and the arrival of the growing season, there is as much as twenty days difference in the fields around the lower and the upper ends of Lake Michigan.

As we continued south we crossed other invisible boundary lines. We left behind the range of the snowshoe rabbit, the star-nosed mole and the short-tailed weasel. Haying increased.

For miles at a time we were once more among hayfields with the cut grass drying in the sunshine, the air filled with its perfume. We, too, were in our haying season, curing memories in the sun of these summer days. Everywhere in the fields there was the sound of tractors. They pulled the mowing machines. They pulled the rakes, the bailers, the hayracks. Along the way, somewhere, we passed a barnyard and Nellie exclaimed: "Why, there's a horse!" Only then did we realize how long it had been since we had seen one in this land of wheels and motors.

I cannot say I regret the passing of the horse as an animal of farm work. The horse-and-buggy age was marred by too much cruelty. Driving on, I recalled stories of early days when teams struggled to drag lumber wagons hub-deep through the mud and bystanders brought clubs, not to help in prying out the wagons but to use in beating the straining animals. I remembered the reply Thomas Hardy received when he remonstrated with an English carter for lashing his mare without mercy: "But she bain't a Christian." Tractors run into ditches. They tip over. They break down. But they are not living creatures capable of suffering as we suffer.

Late in the afternoon when we had almost reached the Kalamazoo River—noting as we went how, in the hazy heat, speeding traffic fed dead or stunned insects to bluebirds that lived dangerously along the highway—we passed a sign pointing west to Allegan. That name stirred to life two far-different memories.

The first concerned a dusty volume I once leafed through, a book of legends of the old Northwest that had been published in 1875. Its discolored title page had carried the name of the author, Flavius Josephus Littlejohn, the imprint of the publisher, the Northwestern Bible and Publishing Company, of Allegan, Michigan, and—most memorable of all—the quaint subtitle the author had chosen for his work: "A cluster of Unpublished Waifs, Gleaned Along the Uncertain, Misty

Line Dividing Traditional from Historic Times."

The second memory is associated with a noble man who died in a small cabin near Allegan during the bitter winter of 1899. He was Simon Pokagon, Indian naturalist, conservationist, author, orator and last chief of the Potawatami. Born in the valley of the St. Joseph River about six miles north and west of the present city of South Bend, Indiana, Pokagon knew only his native language until he was twelve. Later he attended Oberlin College and Notre Dame University, gaining fame for his eloquence. His most widely known book, *Queen of the Woods*, contains poetic passages that mirror his intense feeling for nature. Abraham Lincoln invited Pokagon to the White House and, in 1893, he was a special guest of honor at the World's Columbian Exposition, in Chicago. After we returned home from our summer trip I looked up the oration he delivered on that occasion. A stirring battle-cry for conservation, contrasting the outlook of the white man and the red, it is numbered among the classics of Indian oratory.

"Where these great Columbian show-buildings stretch skyward," Pokagon began, "once stood the red man's wigwam. Here met their old men, young men and maidens. Here blazed their council fires. But now the eagle's eye can find no trace of them. Here was the center of their widespread hunting grounds. All about and beyond the Great Lakes northward roamed vast herds of buffalo that no man could number, while moose, deer and elk were found from ocean to ocean; pigeons, ducks and geese in near bow-shot moved in great clouds through the air, while fish swarmed our streams, lakes and seas close to shore. All were provided by the Great Spirit for our use; we destroyed none except for food and dress. We had plenty and were contented and happy.

"The cyclone of civilization rolled westward; the forests of untold centuries were swept away; streams dried up; lakes fell back from their ancient bounds. Beasts of the field and fowls of the air withered like grass before the flame, were shot for the love of power to kill alone and were left to spoil upon the plains. All our fathers once loved to gaze upon was destroyed, defaced or marred except the sun, moon and starry skies alone which the Great Spirit, in his wisdom, hung beyond their reach."

During his latter years, Pokagon fought to obtain for his people $150,000 owed them by the United States Government. When at last the victory was won, his share barely repaid the expenses he incurred. He died in poverty. His manuscripts were all destroyed in a fire. He lies in an unmarked grave. His name, however, is commemorated, in northern Indiana, by a community of Pokagon and a Pokagon State Park.

The next day, over the Indiana line, we visited the area where his natal village once had stood. No reminder exists today—none except the pleasant land itself. Around us spread rolling green hills, elms and maples, wild grapes and blackberry tangles. Perched on mullein stalks, now in sulphur-yellow bloom, field sparrows spun out the clear trill of their modest songs. All the land lay lush and warm under the sun, the July sun. For now a new page had appeared on the calendar. Now we were in the second month of the second season of the year.

EIGHT

RIVER OF THE FIREFLIES

"THEY soon reached a spot where the oozy, saturated soil quaked beneath their tread. All around were clumps of alder bushes, tufts of rank grass, and pools of glistening water. In the midst a dark and lazy current, which a tall man might bestride, crept twisting like a snake among the reeds and rushes . . . They set their canoes on this thread of water, embarked their baggage and themselves, and pushed down the sluggish current, looking at a little distance like men who sailed on land."

Thus the American historian Francis Parkman describes the events of a December day in 1679 in his *La Salle and the Discovery of the Great West*. On that day La Salle and his followers portaged their eight canoes across a low rise of land from the banks of the St. Joseph River—not far from the birthplace of Chief Pokagon—to a chain of three small ponds southwest of the present city of South Bend, Indiana. The distance was only five miles. Yet the water they left behind flowed toward the Great Lakes, the St. Lawrence and the North Atlantic; the water they reached moved toward the Mississippi and the Gulf of Mexico. They had crossed in that short carrying place the watershed, only a few feet high, that separated two of the great drainage systems of the continent. In the chain of small ponds, the adventurers found the beginning of the dark stream that the Indians called the Theakiki —"Slow River Flowing Through a Wide Marsh"—a name

RIVER OF THE FIREFLIES

corrupted early by the pioneers into the Kankakee.

Two and three-quarters centuries after La Salle, on this day in July, we crossed the Portage Prairie and, where state highways 23 and 123 meet, stopped beside a diminutive pond. One side was marred by an extensive dump. But the other, willow-bordered and rimmed with spatterdock, revealed the original beauty of the spot. This was the last remaining link of the chain of three ponds that, so long before, had led Robert Cavelier, Sieur de La Salle, into the heart of a wilderness empire.

The great marsh through which his canoes threaded their way remained almost unchanged for two centuries—a land of trembling bogs, black mire, cattails and wild rice, waterfowl and will-o'-the-wisps. For more than a million acres it spread away, filling a wide, shallow valley that angles across northern Indiana and over the border into Illinois. The river that flowed through it was so sluggish that the early voyageurs navigated it like a slender lake. They found the ascent of the stream almost as easy as the descent. Early it was called "the river of the two thousand bends." So wandering and aimless was its course that in traversing the ninety-mile length of the valley it traveled 250 miles.

In the days of the Miamis and the Potawatami, the region was a red man's paradise, far-famed for its wild grapes, its teeming waterfowl, its fur-bearing animals. The open prairie west of the St. Joseph sustained such herds of bison that it was known to pioneers as The Cow Pasture. As late as a century ago, the great swamp supplied more than 30,000 pelts of mink, muskrats, otter and other fur-bearers annually. Bass, perch and pickerel swarmed in the dark water. Birds now extinct, the Carolina paroquet and the passenger pigeon, flocked in the region. And during the months of migration waterfowl assembled here in fabulous numbers.

As a small boy in the Indiana dunes just north of the Kankakee, I used to watch skein after skein of Canada geese go honking by, their arrowheads aimed south, over the blue rim of the Valparaiso moraine toward the great marsh that lay beyond. Their destination was only twenty miles away. Yet all through my boyhood it remained unvisited, mysterious, remote, a legendary land of wildness.

To prey on the hosts of the waterfowl, hunters once journeyed each autumn from as far away as New York, Washington and even Europe. They shot until their flat-bottomed boats, piled high with slaughtered birds, sank close to the water line. Then came the meat-hunters, firing endlessly day after day while creaking lumber wagons hauled their kill to the nearest railroad for shipment to the markets of Chicago. These were the so-called "great days" of hunting in the old Kankakee marsh. They extended roughly from 1850 to 1890.

Then, in the early years of the present century, came a sudden transformation. Dredges ripped channels through the black muck of the swamp. They straightened the meandering course of the river, reducing its length from 250 to 100 miles. They cut lateral canals that drained the surrounding marshland. In a decade and a half the river and the marsh, as they had been before La Salle, as they had been before the Miamis, vanished almost completely away. And with them vanished the once-teeming, now-depleted wildlife they had so long sustained.

We tried to picture "The Grand Marsh" as it once had been during the following two days as, swinging back and forth across the shallow valley, we advanced downstream from the source to the mouth of the Kankakee. Into the wide platter of the former swamp we would descend, rising to higher ground on the other side only to repeat the swing in the opposite direction. Each time we would ride for miles across flat land, black land, rich land, the bed of the ancient swamp, a prairie of muck. First in one direction, then in the other, we crossed the river more than a score of times, the amplitude of our swings increasing or decreasing according to the lay of the land. Our route, as we marked it on the map, resembled a thread running in irregular zigzag stitching.

When we first crossed the river, about ten miles from its source, it flowed straight down an excavated canal, walled in with giant ragweed, yellow hedge mustard and elderberry bushes, white with the clumps of their flowers. Six paces carried us across the little bridge that spanned the stream. The water here ran clear and unstained. Downstream the river became, bridge by bridge, more deeply brown, its water darkened by leachings and sediment.

As we stood on that first of many bridges, the morning air was filled with the rich scent of mint. All around us, stretching away in lush rows across the black ground, extended fields of spearmint and peppermint. We were in one of the chief mint centers of the country. Along the upper Kankakee, nearly 19,000 acres produced this crop. Later, as we advanced down the stream, corn and wheat, asparagus and soy beans predominated, with hayfields running along the higher ground at the edges of the valley.

When the great swamp was drained, half a century ago, something like 600,000 acres of the muck-land were considered suitable for agriculture. Now, however, it is recognized that nearly one-fourth of this area is unprofitable to cultivate and may, in time, return again to swampland. Built up by centuries of slow decay, the black, humus-laden soil has special properties. It swells like a sponge after rainfall. In northern Indiana, for example, it is not uncommon for fields of muck-land to carry as much as 50 per cent water.

One farmer in the old Kankakee basin, a decade or so ago, discovered that a five-acre tract of his land was beginning to swell like a giant blister. Year after year the bulge increased. It rose higher than a man's head. It ascended to as much as fifteen feet above the level of the surrounding land. Scientists from Purdue University investigated. They found that the area lay on the edge of an old glacier channel and that seepage from springs kept increasing the water content of the muck. About 70 per cent of the rising land consisted of water. The humus it contained had swelled like a piece of dried apple soaked in water.

From that initial bridge, six paces across, the width of the river increased with each crossing as we progressed downstream. It rose from half a dozen paces to 26, to 30, to 43, to 62, to 101, to 123, to 176 and, finally, to more than 300, almost a fifth of a mile. Great ditches, the Robbins Ditch, the Singleton Ditch—rivers in themselves—joined the Kankakee. Across the Illinois line, beyond the Momence Ledge, that low dam of limestone that long played its part in slowing down the current in the marsh above, we noticed the character of the river changing. It gained speed. It followed a deeper channel. It wound through a land of bobolinks and bur oaks— trees with leaves richly green and glossy above and shining as though plated with silver beneath. In the slower reaches of the stream, islands, often dense with willows, split the current. It was among these islands of the lower Kankakee that botanists discovered one of America's rarest wildflowers, *Iliamna remota*. A member of the mallow family, this bushy plant produces delicately fragrant, rose-colored flowers.

Until we reached the city of Kankakee, in Illinois, the river had dominated all the towns along the way. Here the stream was walled in and subordinate. But, in the open country beyond, the wide and sprawling river became dominant again, the chief object in the landscape. We wound with its advance until, below Channahon, we watched it crawling its final miles close to its juncture with the Des Plaines where the Illinois River is born.

Thinking back on those days with a river, on our crisscross trail down the expanse of the once-great marsh, we remembered varied sounds—some small like the hiss of the current around a trailing grapevine, others loud like the ringing *querr* of a red-headed woodpecker clinging to a dead stub on the

RIVER OF THE FIREFLIES

riverbank. But the recollection that comes back with greatest clarity concerns a sight rather than a sound, an event of the night rather than of the day. It is a remembrance of silence and darkness and the moving, glittering magic-show of the fireflies.

In many places in the course of our wanderings through summer we encountered the dance of these winged lanterns. But nowhere else, nowhere else in all our lives, did we see them in such multitudes as filled the dusk of these nights along the Kankakee. Always this dark stream will remain in our minds as the River of the Fireflies.

There are more than forty orders of the animal kingdom and two of the plant world that possess the power of producing light. Of them all the most famous is the firefly. Two thousand or more species of these luminous beetles have been described in the world. Most of them inhabit the tropics. About fifty species live in the United States. The peak of their display in North America usually comes late in June and early in July. It is then that the galaxies of their winking lights, tracing innumerable glowing lines on the dark, rise across meadows and marshlands to form one of the magical ingredients of the summer.

Somewhere along our way, on some nameless country road, we passed a hayfield at sunset. The windrows curved away like brown rollers in a surf of sun-dried grass, each shot through with shadings of tan and gold and yellow-green. Redwings rode on the crests of these windrow-waves while grackles, hunting crickets and grasshoppers, investigated caverns in the hay. The air, resting at the end of the day, lay calm, redolent with the early-summer perfume of the drying grass. About us, all across the countryside, the whistle of the meadowlark, the jingle of the bobolink, the last song of the day for robin and vesper sparrow, carried far through the quiet air.

Later we came by this field again. The cerise glow had faded from the sky and the deep purple of twilight was merging with the velvet blackness of the night. Birds had fallen silent. The rolling waves of the windrows now stretched away unseen. The beauty of the day was gone. But the beauty of the night had replaced it. For, from end to end, the field was spangled with winking, dancing lights. They rose and fell. They flashed on and off. They waxed and waned in brilliance. At this same moment, over hundreds of square miles around us, this eerie beauty of the fairy-dance of *Wah-Wah-Taysee*, Hiawatha's little firefly, was part of the summer night.

For a long time after we had left this field behind, we followed firefly roads. We made turns, passed dark barns, went by lonely farmhouses where moths fluttered at the lighted window screens. Around us always, wherever we went, streamed the sparks of living fire. We saw them twinkling over the roadside vegetation, above the fields of grain, in the blackness of maple shade, amid the mistily white, faintly seen, flower masses of the elderberry bushes. They passed us in a constant meteor shower. Ahead of us, the twin beams of the car revealed their swarming forms, their lights extinguished by the glare.

Time after time we stopped, switched off our headlights and sat silent, entranced by the scene around us. One whole tree was decorated from top to bottom, filled with firefly lanterns, its dark silhouette sparkling with a hundred moving lights. Beyond, across a lowland stretch, a row of long-dead willows, witch shapes rising out of the darkness, lifted their twisted limbs against the shimmer of the fireflies. So we wandered—half-lost and forgetful of time. For hours we followed little roads, roads without a name, roads we could never find again, but roads we will never forget.

The dance of the lights around us was part of nature's eternal employment of what appeals to us as beauty for the attainment of her ends. The beauty of plumage and song among birds, the beauty of petal color and perfume among

MAYFLIES clinging to the trunk of a tree. With the coming of evening the insects rise in their aerial mating dance.

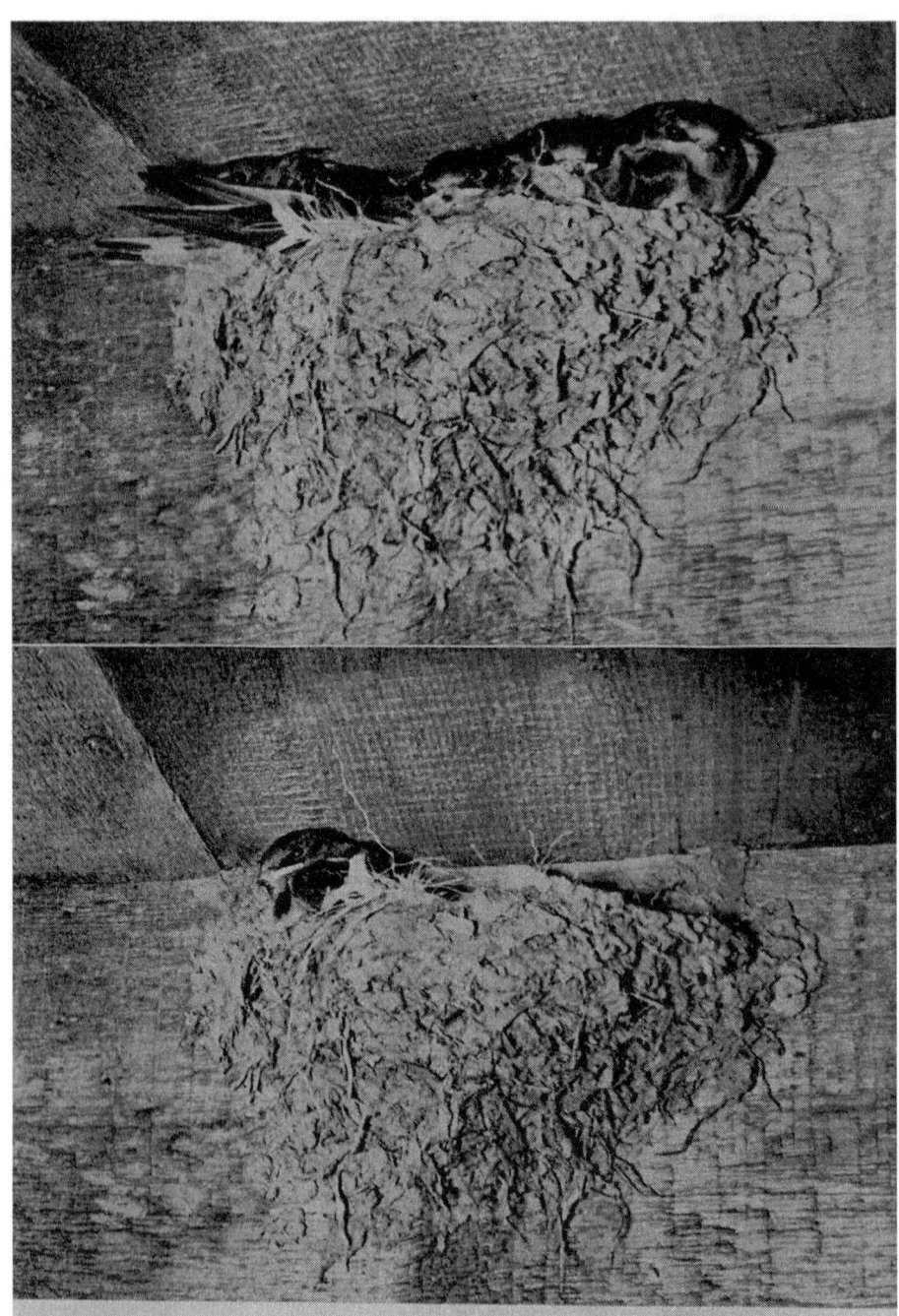

BARN SWALLOWS almost ready to leave the nest. Above, the brood crowded together. Below, the last nestling to leave.

flowers, the beauty of these living sparks of fire all play essential roles in the reproduction of the species. It is through light signals flashed back and forth in the dark that the male firefly discovers his mate. The little lamps of the various species range in intensity from 1/50 to 1/1,600 of a candlepower. Usually the light emitted is greenish-yellow. However it may range from rich bluish-green to beautiful orange-red or reddish-gold. The higher the temperature of the air, the more frequent the rhythmic flashing becomes. Experiments reported in W. V. Balduf's *The Bionomics of Entomophagous Insects* showed that one American firefly flashed, on an average, 8.1 times a minute when the thermometer stood at 66.9 degrees F., while the rate of its flashing increased to nearly twice as fast—15.4 times a minute—when the mercury rose to 83.8 degrees.

Ordinarily the female fireflies have smaller, dimmer lights while the males, to compensate for this, tend to have larger eyes. Those of *Photinus pyralis*, a common species in the United States, are able to see the longest wave length yet recorded for any insect. The number of times the flash of light has to be repeated in guiding the male to the female varies considerably. During laboratory experiments with one American firefly, scientists found the number ran between five and ten.

In all cases the light is produced by the rear segments of the abdomen. Here the outer skin, translucent and unpigmented, forms a window. Beneath it lies a layer of light-producing cells and below that a layer of reflector cells. Within the first layer a substance called luciferin, a cell product, is burned in the presence of an enzyme known as luciferase to achieve what science long has failed to attain—cold light. At least 98 per cent of the energy employed by the firefly is transformed into light. In contrast the ordinary electric bulb dissipates about 70 per cent of its energy in heat. So slight is the warmth produced by the lamp of the firefly that it has been calculated it would take more than 80,000 luminous beetles of one of the largest species to equal the heat of one small candle flame.

By controlling the supply of oxygen to the light cells, the nervous system of the insect turns on and off the light. Different types of flashes—a different light code—is employed by various species. Sometimes, as we saw in Michigan, the lights bob or shimmer in the air. Again, as we observed in northern Wisconsin, they flash on and off, winking or blinking in sequence. Still farther north, in Minnesota, they seem to float, drifting along a few feet above the ground. Here along the Kankakee the predominant motion was upward. Like greenish sparks the tiny lights rose until they were extinguished. The color of the light, its intensity and duration, the interval between the flashes and the number of flashes given in sequence, all these distinguish the signals used by a given species. Each kind of firefly appears to be tuned in to its own combination. In the case of *Photinus pyralis*, the flash of the female invariably follows that of the male by almost exactly the same time—two seconds—in each rhythmic signal and response.

Rhythms infinitely varied run through nature. The famous floral clock of Linnaeus enabled the scientist to tell the hour of the day merely by glancing out his window and noting which plants had opened their flowers. Just as blooms tend to unfold their petals at different hours of the day, so many species of birds begin and end their singing in predictable daily rhythms. Various mosquitoes suck blood at particular hours of the night. Phil Rau, the Missouri entomologist, found that the great American silk moths, the Luna, the Polyphemus and the Cecropia, tend to be most active at specific times during the hours of darkness. So, too, fireflies of varying species begin to flash at different times after the sun sets. Each appears from under the vegetation where it has hidden during the day and commences its evening dis-

RIVER OF THE FIREFLIES

play when the dusk attains a certain density. Night after night, as the failing light arrives at this particular point in its gradual eclipse, the flashing fairy-show of this species begins. Once, when storm-clouds hastened the dusk, we saw the lights of the fireflies come on at an earlier hour. The density that triggered the performance had come before its normal time.

Somewhere along our way we came to a wide expanse of pasture land with a streamlet wandering down its length. It spread away, eerily beautiful, with thousands upon thousands of fireflies lighting up the dark in a maze of moving sparks. They gleamed, fell dark, gleamed again, the number visible varying continually. At times a sparkling wave would sweep far across the pasture as though all the insects were turning on their lights virtually in unison. Then their disorganized flashing would commence again.

We watched closely, hoping to witness a rare and spectacular phenomenon we had heard of but never had seen, the synchronized flashing of fireflies. Under certain circumstances, still not completely explained, all the insects in an area will flash their lights on and off together. Across acres of darkness the night will glitter, then go black, then glitter again, all the lights pulsing in unison, going on and off in perfect synchronization.

A generation ago, scientific publications here and abroad were filled with the great debate over whether such firefly displays existed—and, if they did, what rare conditions produced them. Looking back over the columns of type printed then, a modern observer is likely to conclude that those who had seen the phenomenon believed staunchly in its existence, those that had not seen it doubted it. In this pro-and-con debate the pros have long since won the argument. It is recognized now that at certain times, most often in the tropics, mass flashing does occur.

The earliest record of this phenomenon probably appears in Engelbert Kaempfer's *The History of Japan*, published in

1727. During a trip on the Nienan River, near Bangkok, Thailand, he witnessed a vast display of fireflies, the myriad lights among the trees along the bank winking for a quarter of a mile in simultaneous flashings. In 1938, more than 200 years after Kaempfer, the many ideas that had been advanced in regard to simultaneous flashing were summarized in the *Quarterly Review of Biology* by Dr. John Bonner Buck, of Johns Hopkins University, in Maryland.

One theory held that it was all an illusion, another that it was merely accidental, a third that it was a product of the twitching of the eyelids of the observer, a fourth that it was in some way connected with the sap of the trees on which the insects rested, a fifth that puffs of wind caused the simultaneous appearance of many lights. Other hypotheses suggested that the explanation lay in exceptional conditions of humidity, temperature, air currents and darkness, in the crowding of the insects, in a "sympathy" or sense of rhythm among the fireflies, in the alternate discharge and recovery of "some battery-like mechanism." Still another conception was that the display was produced by a leader, or "pacemaker," whose flash was so closely followed by the insects around that they all appeared to light up simultaneously.

Dr. Buck found that he could make fifteen or twenty males converge on his flashlight by snapping its beam on and off with the same time-interval used by the female. Another experimenter, W. N. Hess, near Ithaca, New York, in 1920, reported he produced synchronized flashing all across a little valley by a similar use of a pocket flashlight. Oftentimes it has been noticed that several males will be stimulated by the flashing of the same female. Thus for a time their lights may become synchronized. On the island of Jamaica, Gerrit S. Miller, Jr., of the United States National Museum, in Washington, D.C., observed clusters of thirty or forty fireflies flashing together. As each luminous insect tends to be set off

RIVER OF THE FIREFLIES

by any nearby light, apparently flashing a little ahead of time as a result of such stimulus, there is always a general tendency toward synchronization. The first firefly to flash after an interval of darkness becomes the pacemaker, speeding up the others around it. With each insect taking its cue from its neighbor in this manner, Dr. Buck believes, the mass flashing is really a very rapid chain of overlapping flashes that, to our eyes, appears simultaneous. Partial synchronization, in which a small group of insects fall into step for a short time, occurs not infrequently. We saw it several times that night across the lowland meadow. But the great mass flashings with tens of thousands of lights coming and going in waves, extending across acres of darkness, winking on and off in unison minute after minute, these occur rarely and under conditions that are still imperfectly understood. It is a spectacle that, in America, only a few fortunate individuals have witnessed.

The second time we saw darkness settle over the wide valley of the drained marsh we were close to the Illinois line, near Lake Village and the beautiful Kankakee State Park and Forest. Here, once more, we wandered for hours amid a carnival of little lights. If possible, the numbers around us were even greater than we had seen the night before. A week or so later and 250 miles to the north, at Sturgeon Bay on the Door Peninsula of Wisconsin, I talked to a truck driver who had circled the lower end of Lake Michigan on this same night. All across northern Indiana, he said, he was driving through a blaze of fireflies. Never before had he seen them so numerous. They alighted on his windshield until all the glass seemed covered with greenish fire.

Wherever we looked that night the lights of the beetles streamed upward out of the darkened vegetation. Observing such a scene near Milan, Italy, in 1824, Thomas Lovell Beddoes wrote home to England that it seemed to him "as if the swift whirling of the earth struck fire out of the black atmosphere; as if the winds were being set upon that planetary grindstone, and gave out such momentary sparks from their edges."

Although the great swamp around us was ditched and tamed, the fireflies spoke of its former wildness. We gave up all attempts to estimate how many hundreds of thousands of insect lamps we saw that night. They trailed above the wild roses of the roadside, across cut-over clover fields that perfumed the soft night air, against the black backdrop of stretches of woodland, above their own reflected lights along the water-mirrors of the drainage ditches. They seemed most numerous of all above fields of oats that stretched away into the night, featureless plains of blackness, but in the day stood out two-toned—rich green among the lower stems, pale yellow-white among the ripening seed-heads.

After the earliest hours of the night, the fever of the fireflies seemed to abate. Their numbers diminished. Each successful mating in the dark extinguishes the spark of the female, subtracts a light from the multitude. Her fertilized eggs are laid in or on the ground, often among mosses or at the bases of moisture-loving plants. The developing larvae within them sometimes emit light even before they hatch. Predaceous, with sharp, sickle-shaped jaws, the immature beetles prey on snails, slugs and earthworms, often subduing creatures larger than themselves by injecting into their bodies a paralyzing fluid. The larval life of many American fireflies extends over a two-year period. Like the mayflies of Kelleys Island, the fireflies of the Kankakee were in their latter days. They, too, were rounding out their lives with an aerial dance. It is believed that most adult fireflies do not eat at all. In their dance of light each generation approaches the end.

From time to time one of the insects alighted on our windshield, tracing glowing curlicues as it wandered about over the glass. Others entered the open windows to circle round and round like shining stars within the space of the moving car. Once we went by a pasture field where black-and-white cows, vaguely seen, were browsing. In the stillness of the night we could hear them cropping, pulling away mouthfuls of grass as they fed. Three or four of the animals were bending down to drink at a small pond or puddle near the fence. About them swirled the gleams of the fireflies and all around their lowered heads the surface of the pool was sprinkled with a hundred glittering, moving points of reflected light. Somewhere, far above the field, above all the surging miles on miles of winged sparks of heatless light, a killdeer speeding through the darkened sky repeated three times its wild and lonely call.

And so we came home out of the wondrous night to the only lodgings we could find, a hot and almost airless room beside that north-and-south artery of truck traffic, U.S. 41. There we lay awake far into the night while the volleying cavalcade roared and thundered by outside. But we were content, remembering the fireflies. As I lay there the words of the old Ojibway chant kept running through my mind: "Fluttering white-fire insects! Wavering small-fire beasts! Wave little stars about my bed! Weave little stars into my sleep!" And when at last we drowsed I have little doubt that memory wove through our slumber, too, images of these little stars, the multitudinous fireflies we had seen.

NINE

WOODPECKER BLOCKADE

THE fireflies of the Kankakee clung to vegetation that now lay 150 miles behind us. We had worked north, first among gravel pits and quarries, then over Illinois flat land, cereal land, with misty trees far across the prairie. We had encountered our first western meadowlark and at an intersection of superhighways beyond the suburbs of Chicago we had observed the complications of modern life epitomized in a traffic sign: "To Make a Left Turn Make Two Right Turns." Corn and wheat diminished as we rode north. Barns became bigger, cows more numerous in rolling dairy land. And on the crests of the low hills the wind swept around us honey-rich with clover perfume or laden with the scent of new-cut hay.

It was now late on the afternoon of the second day. We were seventy miles above the Illinois border, beside one of Wisconsin's innumerable small lakes. Standing there in the sunset we were the lone observers of a prolonged and singular conflict.

Twenty feet above us, a four-foot stub projected from the trunk of a dying oak tree. Twin round holes penetrated this stub two feet or so apart. From the circular doorway of the upper opening three gray heads projected. Framed in the lower hole was the crimson plumage of a red-headed woodpecker. With head outthrust it followed the progress of a parent starling hopping toward the outer twigs on a branch of a neighboring oak. Following a zigzag course through the foliage, the dark bird worked steadily nearer to the upper hole. A large green grasshopper was clutched crosswise in its bill.

The trilling of the three hungry nestlings rose in volume. There was a moment's pause. Then the parent bird launched across the open space. With a flash of red and black and white, the woodpecker catapulted from the lower hole. In a squawking, twisting, fluttering melee, pursuer and pursued disappeared among the branches of the adjoining oak. Then the woodpecker flashed back to the stub, alighted, looked into the lower hole, glanced around, disappeared headfirst and a moment later reappeared in the doorway again. Here it remained, alert, on guard. Within the shade of the nearby foliage we could glimpse the starling hopping sidewise along the branches, approaching stealthily, getting set for another attempt to run the woodpecker blockade.

What had happened was obvious. A pair of starlings had taken over the upper nesting hole after the woodpeckers had excavated it. The latter birds, after chiseling out another hole lower down, were engaged in a constant warfare on the usurpers. Judging by the size of the young starlings, this running battle had been in progress for more than two weeks. In spite of it the nestlings had been brought almost to the point where they could leave the nest. Only a day or two more and they would be in the air. We had arrived during the last act of the drama. This was the critical time. It was now that the growing birds required the most food of all.

During that evening and on the day that followed we watched this avian struggle with unabated fascination. While the sunset glow faded, while rough-winged swallows coursed overhead, while fish leaped and splashed in the lake, the starlings tried endlessly to run the blockade with food. They failed and failed and tried again, they succeeded and failed again. Sometimes there would be a full quarter of an hour's lull in the activity. The woodpecker in the hole seemed to doze, pensive, lost in reverie, as though unaware of what was happening. Then the starling would hurl itself once more toward the outstretched mouths and the woodpecker would shoot into action.

Seven times in succession we watched one of the parent starlings try. Six times it was intercepted and sent fleeing pell-mell back to the adjoining oak. Then on the seventh attempt it came in fast, clung momentarily with fluttering wings and crammed a large white grub into a waiting throat. Sometimes grasshoppers, more often caterpillars, frequently several held in one billful, provided the fledgling with a substantial meal whenever a feeding was successful. Always the transfer of food was a precipitate affair. Once for fully half a minute the long legs of a green grasshopper extended out from the mouth of a fledgling as it struggled to swallow the large insect that had been pushed hastily part way down its throat.

Most of the time it was the red-headed woodpecker waiting in the lower cavity that darted out to repel the starling. But on occasions its mate would come rushing in to intercept a feeding. Then again both birds would join in the attack, their red and white and black plumage flashing in and out among the dark green of the oak leaves. Almost always they observed a curious rule of avian warfare. Their pursuit and attack ceased as soon as the starlings landed. At the end of each swift chase that put the starlings to rout the victorious red-headed woodpeckers would utter a ringing, exultant "Querr! Querr!" that carried far along the shore of the lake. Then, as though to celebrate their success, they would fly away for several minutes and leave the stub unguarded. This was the weak link in their blockade. The starlings always made the most of these short intervals, rushing in for quick, uninterrupted feedings.

On most attempts, when one of the woodpeckers was on guard, the starlings tried to break through by launching themselves straight across the open space from the nearest limb of the other oak. Failure greeted the majority of these attempts. Most birds have one-track minds; they tend to do things the same way over and over. But the starlings, after several thwarted tries, would vary their tactics. Their intelligence and resourcefulness in this desperate game they were playing was the most memorable feature of the long running battle we observed.

After a series of repulses, one of the starlings would sometimes fly to a higher limb and dart down from above, taking the guardian woodpecker by surprise. Again we saw the birds flutter up almost straight from the ground to hang for an instant and make a fast transfer of food at the mouth of the cavity. Once for nearly half an hour we saw the starlings driven off at every approach. Again and again we watched one of the parent birds sidle along the twigs among the foliage while the young birds, yellow mouths agape, gray throats swelling as they trilled their hunger call, waited impatiently. Then came the sudden rush, the attack, the flight, the failure of another attempt. It was at the end of this period that we witnessed an entirely new procedure. It reversed the whole activity of the past.

Leaving the screen of foliage in the adjoining oak, a starling flew by a roundabout course to the rear of the stub and alighted on top. Here it was hidden from the waiting woodpecker. The opening of the nest hole lay almost half way around the stub, just beyond the reach of the parent bird. Calling repeatedly, and each time receiving a crescendo of trilling from the hungry fledglings, it drew them out as far as they could extend their gaping mouths. Then bracing its feet on the bark of a projecting knot, the adult starling leaned farther and farther down until it was just able to make contact and transfer its load of insect food. Too late the woodpecker realized what had occurred and rushed to the attack. The stratagem had worked. The starling once more had run the blockade.

And it had done it by a sudden shift, a dramatic change in tactics. Previously it had fluttered at the entrance of the hole.

This time it alighted and reached down from above. Previously it had worked at top speed to complete its mission before the woodpecker could drive it away. This time it reversed itself, moving cautiously, slowly, deliberately. It substituted stealth for swiftness. It made no sudden movement that would arouse the enemy. It leaned down several times, without haste, slowly drawing out the little birds until they were stretching to the limit. Only then did it attempt to feed. Before the woodpecker could hurl itself from its hole, the fledgling had received its meal and the parent was darting away.

When we left the lake that night, sometime after nine o'clock, the starlings were still winning in this long nip-and-tuck battle to bring off their brood. They were winning it when we returned again early in the morning. And they had it virtually won, their fledglings almost at the time of nest-leaving, when we turned away and headed north later that day.

The warfare of the woodpeckers was against the adult starlings. They hardly seemed to notice the three trilling fledglings leaning far out of the mouth of the nest cavity. Hitching themselves up the stub, the red-headed birds frequently passed close to the nestlings. At rare intervals they would pause and peer curiously at them, much in the same manner that one of the young birds turned its head and gazed intently at a fly that alighted on the bark nearby. Only once during all the time we watched them did we see the woodpeckers make any real attack on the birds in the nest. On this occasion, excited by a futile pursuit, the red-head swooped back to the stub and alighted just above the nest. Instantly the three heads popped out, the yellow mouths gaped, the trilling became a shrill clamor. With three hard, sudden whacks of its heavy bill the woodpecker struck at the open mouths. The heads of the fledglings disappeared. The trilling changed to anguished squawks. Yet only the fraction of a minute went by before the heads shot out again, mouths open, the hunger call redoubled, the young birds apparently none the worse for the assault.

It seemed incredible that the powerful chisel-bill of the woodpecker, able to drive its way into wood, had produced no permanent damage. At this time in the life of a young bird its mouth skin has unusual resiliency and stretching power. It is also exceptionally strong.

On another occasion a woodpecker rushed toward the young birds as though in an onslaught. But at the last moment a curious transformation occurred. Instead of striking at the open mouths with hard strokes, it poked its bill without force in what seemed to be feeding motions rather than an assault. This movement was repeated several times. Was the woodpecker stimulated to such action by the open mouths of the hungry nestlings?

Birds are thus sometimes pulled in two directions. Once, years ago, Dr. Arthur A. Allen, at Cornell University, showed me two robin nests that had been built in the same tree only a few feet apart. The eggs in one nest hatched before those in the other. When the nestlings began calling and stretching up their wide-gaping mouths, the robin that was still incubating her eggs abandoned her own nest and began helping to feed the small birds already hatched. The instinctive urge to feed the young was, for her, stronger than the instinctive urge to incubate her eggs.

Even birds of far different species are sometimes stimulated to feed nestlings not their own. In his *Life Histories of North American Woodpeckers*, Arthur Cleveland Bent reports a case in which a screech owl—after its own nest in the same tree had been broken up twice—tried to adopt and feed a family of young flickers. It brought to the nest pieces of meat which it tore into small bits as though feeding baby owls. In spite of these abnormal attentions, the young flickers, receiving at the same time regular food from their parents, became fledged and left the nest.

Another case of the kind was observed by William M. Lott,

in Ontario, Canada. Writing in *The Canadian Field Naturalist* for April 1939, he tells of seeing a flicker fly to a six-foot stub on May 24, 1938, near Dorchester. As he watched, the bird brought food to the nest and carried excrement away. This it did several times. On examining the nesting-hole, Lott found it contained three young birds. But they were not baby flickers. They were baby starlings.

Recalling such curious facets of bird behavior, we speculated on what we had seen. As we drove away our minds reverted to other woodpeckers, to a strange experience of our own, to a previous June when we, ourselves, had become substitute parents for the homeless nestlings of a family of flickers. In the last act of their little drama, these baby birds afforded us a tantalizing glimpse into an unexplored recess of avian instinct.

Early on the last day of the month we had heard the high whine of a power saw. Workmen were felling a half-dead silver maple in a neighbor's backyard. Only after one of the main branches had crashed to the ground from a height of twenty feet, did the men discover it contained a nest of baby flickers. One was killed by the fall, another died within a few minutes, but the remaining three seemed uninjured. We took the trio in. By nightfall only a blank space and a low stump remained where the tree had risen in which the parent flickers had excavated their nest and laid their eggs and fed their nestlings. For days after the limb came crashing down, the two birds flew from tree to tree calling endlessly, once more than 100 times in succession.

During the week that followed we struggled to feed and raise the nestlings. Only a few pinfeathers showed on their naked bodies. We installed them in a small box set deep in a larger carton, the nearest approximation to a hole we could improvise. Here the little flickers were sheltered and shaded. Here they huddled together, popping up their heads, lengthening their necks, chattering their buzzy hunger-calls each time we removed the porous cloth covering the larger carton to make a feeding. The normal diet of flickers is ants. As many as 5,000 have been found in the stomach of a single adult. Down the throats of the nestlings the parents inject almost a steady stream of this food. Our little flickers were ravenous at all hours of the day. We could not hope to pick up enough ants to meet the demand. What could we substitute? We have since learned that such nestlings thrive on egg beaten up with a little pablum added and the mixture administered with a medicine dropper, two droppers-full each hour. But then we were without experience, confronted by a sudden emergency. We consulted books, called friends, turned to the Audubon Society and the Zoological Society. The majority of those we questioned recommended pellets of dog food. Some suggested we add a little suet, others a little bread, others a little hard-boiled egg yolk, others a little cod-liver oil.

At the end of a slender paddle of smooth wood, we began thrusting the elongated pellets of dog-food mixture into the throats of the nestlings, feeding them every half hour. With stomachs full, they quieted down, occasionally giving a long-drawn call suggesting that made by a chicken in hot weather. Just before they fell asleep they were wont to produce small chirping sounds that resembled the calling of katydids heard far away.

Day by day the pinfeathers increased and elongated. Tails and wings became more obvious. The bills seemed growing harder. At times the little birds stretched their unfledged wings or made small preening motions with their bills among the pinfeathers. They climbed higher on the slanting sticks we placed beside them. Often they cuddled close together in one corner of the smaller box, clinging to the rim.

And so, for four days, all went well. The three little flickers became personalities. Their fate became more and more our concern. In so short a time they won our deep affection. As their hunger grew we fed them as many as 100 times a day.

They never seemed satisfied. It gradually became apparent that the food we fed them lacked something their systems needed. We added a little milk, a little egg yolk, a little cod-liver oil, a little wheat germ. In mounting desperation we tried everything that anyone could suggest. But one by one the nestlings grew weaker. Their chattering hunger calls became fainter. They lifted their open mouths to be fed more and more slowly. With heavy hearts we watched each little bird die quietly, unobtrusively, in the end fading away almost as gently as the melting of a snowflake.

It was on the morning of the Fourth of July, at 6 A.M., that I found the smallest of the three nestlings out of the little box and on the bottom of the larger carton. Never before had any of the three tried to leave the nest box. Three and a half hours later, at 9:30 A.M., it was dead. That afternoon, at 4:15 P.M., I discovered the second smaller flicker out on the bottom of the larger container. Its end came quietly at about eight o'clock that night, three and three-quarters hours afterwards. Only the largest and strongest of the three nestlings remained. Two days later, at noon on the sixth of July, I found this lone survivor out of the little box. I put it back. A few minutes later when I looked again it was out on the bottom of the larger carton once more. At 3:15 P.M. it was faintly breathing but at 3:30—three and a half hours after I found it out of the nest box—life had left this last of the three little flickers.

Looking back we could not help but be struck by this recurring sequence of events. None of the baby woodpeckers ever left the smaller box except just before its death. Then, one by one, they pulled themselves up and out of the nest box. And they did this in each case at about the same time—between three and four hours before the end. One instance of the kind might be an accident, two might be a coincidence, but a three-times repetition of this pattern of events could not be fortuitous.

It occurred to us that we had observed the operation of instinct in a form that, so far as I can discover, has not been reported before. A bird as large as a young flicker would be difficult for the parents to drag from its deep nesting hole after death. If not removed, its putrefaction might kill the other nestlings. So, when they feel the end approaching, such young birds may well have an instinct to climb up and out of the nest themselves. Physiologists know that when death comes gradually there is a point when the process advances beyond the possibility of recovery. At this "point of no return," while their strength is still sufficient for them to leave the nest, some physiological change in the bodies of the young birds may trigger their instinctive act. In the case of each of the three little flickers this point was reached apparently between three and four hours before death.

We had observed these events, it is true, under unnatural conditions. But how could it be observed otherwise? Ordinarily when we find a dead bird under a nest in the out-of-doors we assume it has fallen and been killed or has died in the nest and has been removed by the parents. If we come upon a dead woodpecker beneath a nesting tree we have no way of being sure how it got there. Moreover, dead things in the wild soon disappear, consumed by scavengers. Even if, by a miraculous, billion-to-one chance, we should be passing by just as a fledgling flicker pulled itself out of the hole we would be seeing only an isolated instance that might represent merely an abnormal situation. But in our two boxes we had seen each of the three nestlings repeat the same act at different times but always at almost the same point before the end of its life. It was this repetition that brought conviction. That was the legacy of the short and tragic lives of the three small flickers.

The only corroborative evidence that I have encountered, obtained under natural conditions, was an experience related to me by James Cope, head of the Earlham College Museum, at Richmond, Indiana. Once, while investigating the nest of a downy woodpecker some fifteen feet above the ground, he discovered a dead and decaying fledgling within the cavity. However it was not at the bottom where the other nestlings huddled. Instead it was anchored to the wood of one side of the hole about half way to the top. It appeared to have been climbing up toward the entrance when death overtook it.

As we rode toward the Lake Michigan shore at the end of that second day of observing the woodpecker blockade, we discussed again the significance of the behavior of the three baby flickers. That night I told the woman who owned the cabins where we stopped something of the long struggle between the starlings and the red-headed woodpeckers that we had spent our hours observing. As I turned away I heard an elderly man repairing an electric light ask:

"What does *he* do?"

This question, direct or implied, infinitely varied, was always the same wherever we went. If a man drives a truck or digs a ditch it is solid, simple work. Everyone understands. But if he spends his days roaming afield observing nature or at his typewriter putting down on paper what he has seen, his activity is mysterious, even a bit suspicious. At times when I have explained in answer to this oft-repeated query, the response, half-incredulous, has been:

"What a wonderful way to make a living!"

And to that only one reply has ever come to mind:

"It is, indeed."

TEN

THE ORCHID RIDGES

A GRACKLE went laboring by, trailing a long green pea vine picked up on the highway. Great blue herons, white egrets, black terns, yellow-headed blackbirds, black-crowned night herons passed us tinted in the sunrise. Three and a half hours before breakfast we were watching the awakening life of the great Horicon marshes near Waupun, in southern Wisconsin. This thirteen-mile watery tract, where the Winnebago Indians once gathered their wild rice, now forms one of the most fascinating of all the Federal wildlife refuges of the midwest.

Everywhere in the dawn the multiform life of the wetlands was evident. We watched a muskrat work down into thick soupy mud, its tail floundering about, its hind feet spurting up brown geysers, until at last it emerged, its fur stuccoed and gleaming, to dine on the extracted root of a cattail. Gallinules and rails of two species—Virginia and sora, long bills and short bills together—fed nearby on a mud bank. Swamp sparrows clung to a dozen cattail heads, repeating again and again the sweet little song of their nesting time, their caps and shoulder coloring almost the exact hue of the cattails, their breasts gray like the gray of old spider silk. So, with egrets and rails and gallinules before breakfast, we started our day.

When finally we did eat that morning, it was in unexpected surroundings. We breakfasted in lonely grandeur in the middle of a banquet hall. The only eating place we could discover in Campbellsport was a hotel's expansive dining room. The elderly German woman who acted as cook and waitress came and went among the white-clad tables. As she put down our scrambled eggs, she volunteered:

"Mr. Bauer, as you probably know, died last winter."

We had no idea who Mr. Bauer was. She enlightened us, a sentence or two at a time, as she appeared and disappeared. This had been Mr. Bauer's hotel. He had been a very great celebrity. People used to drive fifty miles from Milwaukee just to patronize his place. He was known everywhere as the biggest bartender on earth. At one time he weighed more than 600 pounds. Even after a long illness, he had still weighed 470 pounds when he died at the age of forty-six.

For a time after Campbellsport had been left behind, as we drove toward the Door Peninsula, our road carried us among low, rounded hills and bowl-shaped depressions. We were traversing the singular kettle moraine country of eastern Wisconsin. Born of the Keewatin Ice Sheet, each kettle marked the place where a pocket of glacial ice had melted and produced a depression.

Beyond a town of Two Rivers and a village of Two Creeks, we angled out onto that long thumb of land that forms the Door Peninsula. French voyageurs, centuries before, had named the dangerous passage between rocky islets at its tip, *Porte des Morts*. Hence the designation of this peninsula which, for more than seventy-five miles, forms the lower shore of Green Bay. It was here, in 1634, that the French explorer Jean Nicolet, discoverer of Lake Michigan, leaped ashore, clad in mandarin robes, firing pistols with both hands, believing he was landing on the coast of China. It was also along this shore that Pere Marquette and Louis Joliet paddled on their way to the Fox River and the discovery of the Mississippi.

As it advances northward, the Door Peninsula grows steadily narrower. The water draws closer on either hand and its moderating influence is reflected in the widespread cherry

THE ORCHID RIDGES

orchards of the region. A paradox is produced. The northern part of the peninsula has a milder climate than the southern part. No other county in Wisconsin, not even those along the southern boundary of the state, has so long a growing period as Door County. It enjoys an annual average of 160 frost-free days. Normally at the oldest village on the peninsula, Baileys Harbor, the lake water remains open all winter long.

It was Baileys Harbor that was our goal that day. For, just beyond it, low ridges, like a succession of parallel waves, extend inland from the shore. In the boggy troughs between, orchids bloom in wild gardens that have made the region famous. Dr. Albert M. Fuller, Curator of Botany at the Milwaukee Public Museum, reports that thirty of Wisconsin's forty-five kinds of native orchids grow in Door County and that twenty-five are found in the small area of The Ridges. In the fall of 1937, these orchid bogs obtained permanent protection as a wildflower preserve when a group of local citizens organized and incorporated The Ridges Sanctuary. The work of this group was honored in 1942 when it received the Garden Club of America's annual award for the most worthy project in conservation.

It was our good fortune to ask for directions at the Baileys Harbor Library. The Librarian, Olivia Traven, we discovered, was one of the signers of the articles of incorporation for the sanctuary. She had known the area intimately for years. No one could have been more helpful than she was when, on the following day, she showed us the wooded ridges and all the unique beauty of this region which, just in time, had been saved from destruction.

The waves of sand, fifteen in number and a couple of miles long, are at most only about six feet high. They extend inland for a mile or more. Apparently they represent the long-ago shallows of the lake edge, for swimmers who wade out from shore find similar ridges beneath the water. At one point they will be only ankle-deep, a little farther on up to their chins, then ankle-deep again. On shore, the successive ridges become older and more heavily wooded as they advance inland. Conifers predominate—hemlock and balsam fir, white pine and black spruce and tamarack. Each ridge has its name: Wintergreen Ridge, Sandy Ridge, Deer Lick Ridge. And each has its own trail running down its length and overlooking the bog strips on either hand. All told, 750 acres, well over a square mile, are included within the boundaries of this wildflower preserve.

The first orchid bloom we saw amid these orchid ridges was the delicately beautiful grass pink. Then came the rattlesnake plantain and, scattered all along the sides of the ridges, the tinted slippers of the pink moccasin flower. The great spectacular orchid show of the year had come and gone during May and June. The two rarest species of the sanctuary, the Calypso and the delicate ram's-head lady's-slipper, were both now past their blooming time. Although we had arrived too late for most of the flowers, orchid plants of many kinds rose around us as we wandered down the ridge paths and along the edges of the bogs.

For here are rooted the rose pogonia, the heart-leaved twayblade, the tall white bog orchid, the tall leafy green orchid, the white adder's-mouth orchid, and the long-bracted orchid. Here grow the large round-leaved orchid, the smaller purple-fringed orchid, and the blunt-leaved orchid.

Long ago, John Gerard, surgeon and herbalist of sixteenth-century England, wrote eloquently of the infinitely varied beauty of the orchids. "There be divers kindes of Fox-Stones," he noted in his famous *Herball*, "differing very much in shape of their leaves, as also in floures: some have floures wherein is to be seen the shape of sundry sorts of living creatures; some the shape and proportions of flies, in others gnats, some bumble bees, others like unto honey bees, some like butterflies, and others like waspes that lie dead; some yellow in color, others white; some purple mixed with red, others of a brown overworne color: the which severally to distinguish, as well those here set downe, as also those that offer themselves darkly to our view and consideration, would require a particular volume, for there is not any plant which doth offer such varietie unto us as these, except the tulipas, which go beyond all account."

Three of the lady's-tresses—the slender, the hooded and the nodding—bloom among the ridges. And here it is also possible to find three of the coralroots—the spotted, the early and the striped; three of the rattlesnake plantains—the giant, the dwarf and the intermediate; and four of the lady's-slippers—the showy, the large yellow, the ram's-head and the pink.

Among the bogs of Wisconsin, the welfare of the lady's-slippers depends, to some extent, on what happens to the rabbits. From 1932 to 1935, Aldo Leopold once pointed out, rabbits were abundant. They browsed heavily on the bushy little bog birches, letting in the sun, opening up the bogs and benefiting the lady's-slippers. Then during the years 1936 and 1937 the rabbits reached the low point in their cycle. The bog birches grew dense and once more deeply shaded the orchids. The following year, 1938, the rabbits recovered, built up their population, mowed down the bog birches. And again the lady's-slippers were thriving in the added sunlight.

As we moved along the ridge-top trails, that day, the sky above us darkened and a quiet rain began to fall, gentle and warm. Here and there, where the ridge slope met the bog, the dwarf lake iris, *Iris lacustris*, a plant with leaves only four or five inches long, spread in dense mats over the ground. Now they were richly green. But in the spring each mat had been a pool of blue. Then the gay-wings, *Polygala paucifolia*, had been in flower and even before that the inner ridges had known the beauty of petal and perfume brought by the trailing arbutus. Around us now the leaves of the pipsissewa, the clintonia, the white-flowered shinleaf were wet and gleaming. In the slow rain, the plants of bog and forest—the red wood lilies, the snowy-headed cottongrass, the marsh anemone—all were washed and sheathed in moisture, all possessed a glowing kind of beauty even in that dull and leaden light.

We found it pleasant there in the dripping woods, the smell of bog and mold rich in the air, the gentle rain falling, the trails leading us on. We stopped beside a decaying log. It was mantled, like the logs of Mt. Mansfield, with a whole miniature forest of lichens and mosses, seedlings and bunchberries. Witch's-brooms, those still mysterious spurtings of tree growth that produce tangled masses of twigs—known in the west as "porcupine nests"—appeared in many of the higher junipers. On the ground below, almost vinelike, the prostrate or trailing juniper, *Juniperus horizontalis*, sprawled in outspread masses. Some of the berries were blue, some still pale green. For this fruit waits a second year to ripen.

Later that afternoon we were joined by Emma Toft, a remarkable woman well along in her seventies who had lived all her life among the wildflowers, the birds and the animals of an area left relatively unchanged, the pioneer homestead where she was born, Toft's Point, just north of Baileys Harbor. A long time ago an Eastern botanist had come to the Door Peninsula to collect plants. He had boarded with her family. Thus as a small child she had become fascinated by the study of wildflowers. It was an interest that never waned, an interest that had played its part in the preservation of the orchid gardens of The Ridges.

Of all we saw that day, the memory that lingers most vividly in our minds is coming suddenly upon almost half a hundred "whip-poor-will's shoes"—showy lady's-slippers—all in bloom in an area no larger than a city lot. Each waxy-white and crimson-tinted flower, delicate as a bubble, hung pendant on its stalk. All around the air was filled with a faint, indescribable perfume. Among these orchids we came upon a number with severed stalks where flowers had been nipped off by deer. The animals, browsing here, seemed to choose such flowers as a special delicacy. Later, on the Upper Peninsula of Michigan, we came upon a pond where deer had fed, apparently by preference, upon the fragrant white blooms of the water lilies. I remembered the Cape Cod child who, Henry Thoreau records in his *Journal*, asked about the bobolink:

"What makes he sing so sweet, mother? Do he eat flowers?"

Delicately formed and gracefully beautiful enough to have been nourished by flowers was the week-old fawn we met when we rode down a dirt road lined with great-leaved thimbleberries to reach the homestead on Toft's Point. Fishermen had discovered it stranded on rocks near the lake's edge. They had taken it to the local game warden who had asked Miss Toft to raise it and then set it free. As we watched, it fed ravenously on warm milk from a bottle. But already it was nosing among leaves, nibbling tentatively here and there. With its spotted sides, its slender neck, its mobile ears, its long, curving eyelashes, its liquid eyes, it seemed at the time the most beautiful of all the world's beautiful and appealing wild creatures.

The next day we saw another wild creature that was appealing, also, in a different way. It was one of the innumerable gallant forms of wildlife that continue courageously, as best they can, in spite of serious handicaps. We had stopped to watch a flock of martins skim and dive, twist and flutter as they fed not far from the lake shore. One after another they settled to rest on telephone wires. Each checked its speed, grasped the wire with both feet, folded its wings—all except one. This swallow, through injury or disease, had one leg that hung shriveled and useless. We watched it dip toward the wire, hover, grasp the support with its one good foot, then, at the last instant, extend its left wing to touch the wire and provide a stabilizing second point of contact. In this position it rested, half lying on the wire, skillfully maintaining its balance in spite of the light breeze that was blowing.

In many ways, in many places, I have seen wild creatures

compensating for their injuries. With fortitude and instinctive courage they continue to survive under handicaps that seem overwhelming. Along the New Jersey coast, we once watched a Husdonian godwit approach us as it fed along the edge of a tidal bay. As it walked its left knee bent more deeply than its right. It drew close and we saw its right leg had been shot away just above the foot. By compensating for this shorter leg in its knee movements, it was getting about and feeding just as it normally would do. Another shore bird lived for at least five years with one leg completely missing. On Laysan Island, in the Pacific, a lighthouse keeper discovered an injured western golden plover. He amputated its leg and later released it. Every winter for five years it returned, across more than 2,000 miles of open water, to the same beach on the same island.

For three years a red-winged blackbird with a single leg came back to the feeding station in our yard on Long Island. Each time it landed we would see it twist and sideslip at the last moment before it made contact with the ground. In this way it killed its forward speed and threw the remaining momentum toward the side of its good leg. In hopping about, as it fed, it leaned slightly to the right to maintain its balance. A male, it remained in exuberant health and spirits all the time we knew it.

In this same Long Island backyard we were once visited for several days by a starling that had lost the sight of one eye. Its right eye was white, opaque, entirely blind. Yet, amazingly, it survived day after day. It flew with the other birds and alighted among the branches of a tree always with its head carried far to the right in order to see ahead. Several times we observed it perform a curious maneuver. Whenever the other birds in the yard flew up in alarm from their feeding, this starling would crouch and whirl about in a swift circle on the ground as it swept the whole surroundings with its one good eye.

For several successive years, we noticed in our neighborhood a strain of starlings that possessed abnormally long bills. One in particular was handicapped by a yellow bill fully half again as long as those on the other starlings around it. It dipped or drooped at the tip like the bill of a curlew. Within it, the tongue, apparently, was of normal length and was unable to reach the food the bird picked up. A hundred times we saw this starling repeat the same procedure. It would grasp a bit of bread with the tip of its bill. Then it would place the side of its head on the ground, sliding the bill along until it had worked the food back within reach of its tongue. Day after day we saw it feed in this manner, maintaining itself, well-nourished and active as long as we observed it.

Infinitely varied are these adjustments that handicapped creatures make. Bob Hines, the noted U.S. Fish and Wildlife artist, once told me of a gray squirrel he watched for several months among the elms of the White House lawn, in Washington. Through some accident it had lost the banner tail that is so important to such squirrels in maintaining balance as they leap from limb to limb high above the ground. This tailless squirrel, he noted, recognized its limitations. When pursued by other squirrels it never leaped from tree to tree. No matter how closely pressed it was, it always raced down the tree trunk and across the lawn to an adjoining tree instead.

At the time he was field naturalist for the Alaska Game Commission, Frank Dufresne once came upon several black bears at the edge of a salmon stream. One of the animals, he told me, had a curious appearance; it seemed to have a beard hanging below its chin. Through his fieldglasses he saw that the bullet of a high-powered rifle had shattered its lower jaw. Most of it hung down completely useless. In spite of this terrible injury the bear was one of the fattest in the group. How could it feed itself? Dufresne watched closely. Several times he saw the animal splash across a small pool and throw itself on salmon it had frightened into the shallows. Ripping chunks from the sides of the fish, the bear stuffed them down its throat with its forepaws, making little if any attempt to chew. The bullet that had struck it had burdened it with what seemed a mortal handicap; yet with wild courage it was surmounting this overwhelming catastrophe. It was getting all the food it needed and was fat enough to hibernate.

Back in the forest of Minnesota, a friend of mine once saw several deer crossing an open space. One had only three legs. Its left foreleg apparently had been shot away. It appeared as sleek and well-fed as the other animals and had no difficulty keeping up with its companions as they disappeared into the woods. Such creatures often get along for a considerable period. But they are always vulnerable in the face of some extra strain, some unusual emergency that confronts them suddenly.

In *Circle of the Seasons*, I told of a blue jay with an injured wing that was unable for several days to ascend in flight. It could navigate the air only on a long downward slant. This bird—and during a later year a second blue jay—would hop from the ground into a bush and from the bush into a tree and up the tree from branch to branch until it reached the top. There it would launch itself out on a gradually descending flight to a tree in another yard. Later in the day it would reverse the process and return to our yard again. We watched the bird repeat this performance day after day until the muscles of its injured wing had regained their strength.

On the other side of the continent, a golden eagle with an injured wing was an object of special interest when, twenty years ago, James B. Dixon reported in *The Condor* his studies of the eagles of San Diego County, in southern California. Over and over he watched with fascination the same stratagem used by an old female after she had been maimed by a blast from the gun of a duck hunter. The lead pellets severely injured the end of one wing. Afterwards the eagle was able to fly only by tilting up the opposite wing and much of her former speed in the air was gone. She compensated for this handicap by means of an ingenious ruse.

THE ORCHID RIDGES 105

Her nest occupied the top of an oak tree on a rough mountainside overlooking a fresh-water lake inhabited by coot. Around the edges of this lake ran a barbed wire fence. For a considerable distance the fence was covered by wild morning-glory vines, forming an almost solid screen. It was the habit of the coot to come out on the lake shore near this section of the fence to feed and sun themselves. The old eagle would leave her nest and sail away high in the sky as though leaving the area behind. Then she would drop down and return, skimming along only about three feet from the ground until she was paralleling the fence and its screen of vines. Opposite the coot she would suddenly lift over the barrier and drop on her unsuspecting prey. Time after time, Dixon saw her repeat this maneuver and always with success.

On the night before we left the Door Peninsula, as we were driving in moonlight, we noticed the speedometer reach and pass the 5,000-mile mark for our summer trip. A little after eight, next morning, we were running south along the Green Bay shore with storm clouds, gray-black and slaty-blue, tumbling, piling up, filling the sky and darkening minute by minute ahead of us. One of those violent thunderstorms that between thirty and forty times a year sweep across this part of Wisconsin was charging toward us.

Around the world, night and day, it is estimated that there are about 1,800 thunderstorms in progress at any given time. On the same world-wide basis, there are about 100 flashes of lightning every second of the year. But the storm that enveloped us that morning provided us with a new experience. It was unlike any that either of us had encountered before.

To our right and left, the tumble of the menacing clouds descended low. Wind-torn vapor, like gray, plucking fingers, continually in motion, raced across the fields. These low-descending clouds formed the abutments of a vast arch that curved upward over the road before us. There the storm vapors were hundreds of feet in the air. The sky seemed lifting to let us pass. Magically, mile after mile, we drove through the arch of the storm.

In flashes, less than a ten-thousandth of a second in duration, lightning ripped through the dark clouds on either hand. One thunderbolt ran zigzag along the horizon to our right. Each stroke hurled the vast explosion of its thunder over us. Yet we seemed less in the midst of the storm than running between two storms. Steadily the darkest clouds, the brightest flashes, the loudest cannonading moved away behind us. We were through the arch and under lighter sky. Then the deluge began.

It came almost without warning. One instant the roads were damp with vapor moisture; the next they were streaming with running water. In a widespread downpour as many as 3,000,000 gallons of water may fall. According to the records of the U.S. Weather Bureau, the heaviest rain ever measured scientifically occurred in California. There 1.02 inches fell in one minute. If this descent had continued, it would have rained sixty-one inches in an hour or 122 feet in a day. When rain falls at the rate of four inches an hour, it is considered a cloudburst. For a time we thought all these things were occurring around us. The windshield wipers were helpless. I crept along, driving almost by feel, peering ahead through the streaming water on the glass and the opaque curtain of the deluge beyond, until we glimpsed a side road. I pulled off and we sat out the rest of the storm.

The drumming of the great drops on the roof of the car echoed in our ears. Falling through still air, a raindrop strikes at about twenty miles an hour. The largest drop that reaches the earth, studies have shown, is 0.3 of an inch in diameter. Larger than that, it flattens out and splits apart in the air. Minute by minute, as we sat there, the pounding of the rain continued unabated. This storm was a goose-drowner, a gully-washer, all the things country people call the heaviest of the summer rains.

But in time the drumming slowed, visibility increased, and we drove on through the lessening rain. Once we passed a side road, a ribbon of reddish mud that reminded us we were close to Red Bank where Nicolet had leaped ashore and where the Indians had welcomed him with a feast of 200 roasted beavers. Behind us the curtain of falling rain moved away over the Door Peninsula. It was washing clean the billion leaves of trees and bushes and herbs. It was flooding ditches, creating barnyard puddles, splashing down on field and swamp and shoreline. And back at The Ridges it was contributing additional moisture to that succession of slender bogs that provide so many varied forms of orchids with a home.

ELEVEN

HIGH ROCKS

THE next day we traveled west and north, inland from the Lake Michigan shore into upper Wisconsin. We entered a region of bracken and leatherleaf, white birch and red pine and mile after mile of aspens that had sprung up in the wake of the saw. In the morning mist little bogs echoed with the slow grating of frogs like the metallic winding of small alarm clocks. We stopped beside the Peshtigo River across which, in 1871, had raced the most destructive forest fire in the history of the United States. Spread by near-tornado winds, the flames had devoured the forest through six counties, had wiped out whole communities and had caused the death of 1,200 persons—all in the space of four hours' time. Later that morning the mist rose, the sun shone and we advanced through all the thousand subtle shadings of Wisconsin green.

Once we came to a small restaurant with a lettered sign in the window: "Wild Rice Pancakes." It brought to mind all the unusual foods we had encountered in our travels—turtle steak, lime pie, goat's-milk fudge, lobster rolls, conch stew, crayfish bisque, cactus-juice milk shakes and salmon chowder. By an easy transition we found ourselves recalling the wonderful wild flavors of the out-of-doors, the taste of pawpaws and thimbleberries, of black walnuts and red thorn apples, of wintergreen leaves and sassafras twigs, of wild strawberries picked in the spring dew and persimmons gathered in the frosts of fall. We stopped that day extra early for lunch.

THE KANKAKEE on a summer day. During the nights of early July, fireflies flash in the darkness all along its banks.

HIGH ROCKS and their collector, George H. Peters, above. Shown below, a firefly clinging to a dewy grass head.

HIGH ROCKS

Sundown found us east of Crandon, near the headwaters of the Peshtigo, following the pleasant path that climbs the western slope of Sugarbush Hill. On our road map and on the lists of the U.S. Geological Survey the summit of this hill carried the designation: "Highest Point in Wisconsin."

The air was calm. Every grass blade we touched along the way added shining motes to the clouds of tiny insects dancing in the late sunshine. White admiral butterflies wandered over the yellow mullein and the red hawkweed. And all through the lighted treetops around us birds darted and sang—scarlet tanagers, chestnut-sided warblers, red-eyed vireos and indigo buntings. A least flycatcher called endlessly from an upper branch. I looked at my wrist watch. Its "Chebec! Chebec! Chebec!" came forty-five times in sixty seconds.

Stopping often to gaze back or to browse on ripe raspberries hanging beside the trail, we ascended to the top. Before we were half way up, the valley below lay in shade and the shadow of the westward ridge was climbing the slope behind us. The smell of balsam strengthened in the evening air. From the dusk of the forest on either hand came the voices of thrushes—the wood thrush, the hermit thrust, the veery. We reached the summit before the shadow and for a time stood in sunshine amid the sugar maples that give the hill its name.

Up this same path, toward this same summit, a friend of ours had climbed but a few years before. When he reached the top he was close to the end of a twenty-year avocation and quest. Already we have crossed his trail a few chapters back at the place where he and his plants were temporarily stranded on the bridge near Niagara Falls. Wherever we would go over the face of the United States, we would encounter and re-encounter the paths of his wanderings. For this Long Island botanist and mountain climber, George H. Peters, had spent two decades of short summer vacations in the pursuit of a prodigious hobby. Traveling some 60,000 miles, he had ascended the highest points of forty-nine of the fifty states, the single exception to date being Alaska. And he had brought home from each summit a bit of rock. Labeled and dated, these high rocks now are ranged on the shelves of a special oak cabinet at his home in Freeport, New York.

The highest stone came from the 14,495-foot peak of Mt. Whitney, in California; the lowest from a knoll-top in western Florida, only 345 feet above sea level. Each rock in the cabinet, each bit of gray limestone or red granite or black volcanic rock, each piece of conglomerate or quartz or schist, recalls to this friend of ours memories of adventures on widely scattered heights.

Most of his life has been spent close to sea level on Long Island, where he is Deputy Commissioner of Public Works for Nassau County. In the course of botanical rambles, during summer vacations, he scaled such eastern peaks as Mt. Katahdin in Maine, Mt. Washington in New Hampshire and Mt. Marcy in New York. It was not until the early 1930's that his avocation took the strenuous form of an attempt to climb to the highest point in as many states as possible.

Year by year the rocks that formed the trophies of his ascents grew in number. Using thumbtacks he located the high points of the various states on a large map of the country, employing red markers for those still unclimbed and blue for those already scaled. In time the blue thumbtacks outnumbered the red. Then they became the overwhelming majority. Finally, in the summer of 1955, Peters pulled the forty-eighth red marker from the map and inserted a blue one in its place. Three years later, in 1958, on a flying trip to the Pacific, he climbed to the summit of the volcano Mauna Kea, highest point in Hawaii. This added a forty-ninth pin to his map and a fragment of black lava to his collection.

On several evenings at his Freeport home, Nellie and I had examined the pin-studded map on his study wall and handled the rocks that had come so far, from such various locations. And as we examined them, George Peters was reminded of the two-score and more climbs that had brought together this collection of hard-won stones.

To reach Michigan's highest point, in the Porcupine Mountains of the Upper Peninsula, he had to swim a forty-foot stream, pushing his clothes on a log before him. In one Southern state when he inquired at a filling station about a road ascending part-way up the peak he wanted to climb, he was urged to circle around the mountain and use another road. The reason: too many moonshiners had stills in the woods on this side of the peak. On the flat plains of western Kansas, his problem was to determine which point really was highest. Again and again Peters would sight along a hand level laid on the ground and decide some spot a quarter of a mile away had a few inches greater elevation. When he had finally convinced himself he had found the right place, he had to hunt for a long time, wandering in expanding circles, before he could find a piece of rock on the almost stoneless prairie.

Indiana's high point, Greenfork Top, is north of Richmond. It proved to be a knoll in a cornfield. Peters stopped at the farmhouse to explain his errand. As he was talking to the young woman who answered his knock, the voice of an older woman came from a back room:

"Who is it?"

"A man who says he has come all the way from New York to get a rock from our cornfield."

"Well, shut the door—quick!"

Even after he had been granted permission, Peters' troubles were not over. Amid the cornstalks that rose higher than his head he could not get his bearings. He had to rely mainly on feel to determine when he had ceased climbing. For half an hour he walked back and forth along the rows before he was sure he had reached the highest point in the field. The elevation of Greenfork Top is 1,240 feet; that of Charles Mound, near Freeport in northern Illinois, nearly 300 miles away, is 1,241 feet. Thus there is a difference of only a single foot between the high points of these two states.

Although it is not one of the highest elevations in the country, Minnesota's loftiest ground proved to be one of the most difficult spots to reach. A dozen miles south of the Canadian line, it lies among the Misquah Hills of the Superior National Forest, the greatest wilderness area in the country. No topographical maps of the region have ever been prepared. It is canoe country, a land of tangled forest and innumerable lakes. When Peters inquired for directions at the ranger station he was instructed how to paddle from one lake to another in a roundabout route to reach the Misquah Hills. He explained he did not have that much time. His vacation lasted only two weeks and he had other mountains to climb. With much head shaking, he was directed down an unused lumber road and told if he had not returned by the second day the rangers would have to start looking for him.

At the end of this primitive road he struck out on a compass course straight through the forest. The date was August 9, 1940. Breathless, sweltering, humid heat enveloped the woods. At times he had to worm his way through tangles of fallen trees. In other places he pushed ahead among ostrich ferns that rose above his head. He came to a forest lake and had to circle the shore and take up his compass line again on the other side. Among old burns he moved ahead through solid stands of raspberries with the ripe fruit falling like a shower of large raindrops all around him. More than once he came upon bear wallows, in one case with the brown water still roiled from the sudden departure of the big animal. He tried to whistle and call as he advanced to warn bears of his approach. But in his dehydrated condition his dry lips and throat produced almost no sound at all. On the way back, after he had found a small clearing on a ridge and the remains of a triangulation tower marking the high point, he suffered increasingly from heat exhaustion. At times the pounding of his headache became so severe he would lie down for a quarter of an hour before he could get up and go on again.

At the opposite extreme was the piercing cold of a high-country blizzard that surrounded him when he made his first attempt to scale the 12,850-foot Granite Peak in the Beartooth Range of western Montana. He started the ascent alone on July 29, 1951. Intermittent flurries of snow surrounded him. At times, as he climbed higher, the falling flakes hung like a curtain, blotting out the valley below. By nightfall he was among the last trees of the timberline. He camped at the edge of a glacier. Encouraged by better weather in the morning, he started on. But by afternoon he was inching his way along the knife-edge of a high ridge with snow swirling around him in a virtual blizzard. Momentarily the snow curtains would sweep aside, revealing the chasms that yawned on either hand.

In a little saddle on this 12,000-foot ridge, almost exhausted and shivering from the blast, Peters realized he would have to turn back. Time had run out. If he was to reach the shelter of timberline by dark he must start immediately. Under this added load of disappointment and defeat, his strength seemed to drain away. Time after time, gasping for breath in the rarefied atmosphere, he struggled to clamber up eight feet of overhanging rock out of the saddle onto the sawtoothed ridge. And time after time he failed. For ten minutes, completely beaten, he lay on the rock of the knife-edge, the snow piling up around him. Then, numb with cold but with some strength regained, he tried again. In desperation he inched himself up the last few feet and gained the ridge top. Then began the seemingly endless battle against cold and exhaustion on the long descent to his previous camp.

Two years later, in August 1953, Peters came back again to Granite Peak. This time the assault on the mountain was carefully planned. Accompanied by Jerry Brandom, of Portland, Oregon, another climber making a try at the formidable peak, Peters rode by pack train to timberline. From there they started fresh. Even with this advantage they barely reached the top. Their long climb took them past the towering ice wall of a hanging glacier, inch by inch up a 200-foot crack in the sheer face of a cliff and through "the eye of the mountain," a hole penetrating a great rock close to the summit. One previous year thirty climbers had tried to attain the pinnacle of Granite Peak. Not one had succeeded. Peters and Brandom were the twentieth and twenty-first persons ever to stand on this forbidding summit. The red piece of rock that Peters brought back to join the only mineralogical collection of its kind in the world is granite similar to that found at the top of Pike's Peak, in Colorado.

The rock he picked up on 8,751-foot Guadalupe Peak, the highest point in Texas, was white and filled with fossils of small sea creatures. The one he found, at an elevation of 12,655 feet, on Arizona's highest mountain, Humphreys Peak, was black and of volcanic origin. Black also was the rock he brought home from the high point of North Dakota. With the toe of his boot, a farmer drew a map in the dust to direct Peters to this eminence, Black Butte. For on that day a low-lying layer of clouds was cutting off the tops of the buttes, giving them all the appearance of rising to the same height.

When Peters reached Arkansas on his Grand Tour of America's High Spots, he found he had not one but two mountains to climb. In the northern part of the state, Magazine Mountain has an elevation of 2,800 feet. To the south, in wild country west of Hot Springs, Old Blue Mountain rises to exactly the same height above sea level. Lying within a state park, Magazine Mountain was easy to ascend. It was Old Blue that proved formidable. Remote, timbered, covered with thorn thickets, reputed to be alive with rattlesnakes, a long ridge ascends to the summit with numerous side ridges joining it during the ascent. Peters made the climb on a day of sultry weather climaxed by a thunderstorm. Much of the time he was pushing through briary tangles so dense it was impossible to see over or through the underbrush. Everything was parchment dry. Dead leaves covered the ground. They fell in a shower whenever he touched a branch. On all sides there was a continual dry rustling as though snakes were everywhere. But he saw none and, using a stand of pawpaws as a landmark, succeeded in keeping to the main ridge on the upward climb and on the almost equally difficult descent.

No thorn-thickets, no blizzards, no difficulties of any kind presented themselves when he rode up to the high point of Delaware. Four hundred and forty-two feet above the sea, it lies on Ebright Road near the Pennsylvania line about six miles north of Wilmington. Across the open top of West Virginia's loftiest land, Spruce Knob, Peters found acres of wild bleeding heart in bloom when he visited the spot on the nineteenth of July, 1939. A dozen years later, in the same month, on July 9, 1951, he stooped to examine the blue-purple blooms of the highest flowering plant in the United States, a small clump of *Polemonium*, the "sky pilot" of the mountains, growing within six feet of the 14,495-foot summit of Mt. Whitney, in California. According to the code of the Sierra Club, any mountaineer climbing above 13,000 feet is entitled to wear a sprig of sky pilot in his hatband. On his way down, Peters plucked several of these flowers. He still has them. They form a special feature of the extensive herbarium of pressed high-country plants he has assembled.

In more than one instance George Peters' hobby has resulted in altering the official records of the Geological Survey, in Washington, D.C. Before World War II the highest point in New Mexico was listed as North Truchas Peak. To reach it Peters climbed a sister mountain, South Truchas Peak. While at the summit he sighted along his hand level and discovered that all the mountains around, including North Truchas Peak, seemed to be lower. Descending into the valley between and climbing the north peak he repeated his hand-level test there. It confirmed the fact that South Truchas Peak, not North Truchas Peak, was really the higher of the two.

When he wrote to the Chief of the Map Information Office of the Geological Survey, he was told that as soon as the war was over a new survey would be made of the area. In the course of time a subsequent letter arrived. It announced that the survey had been completed. It corroborated his findings. South Truchas Peak was higher. As Peters had already collected a rock from each summit he felt safe whichever mountain proved to be the loftier. But, the letter continued, during the re-survey it was discovered that Wheeler Peak, 150 miles away, was forty-one feet higher than South Truchas Peak. It was the real high point in New Mexico. So in the spring of 1955 Peters drove west again, climbed this 13,151-foot mountain and brought home a bit of conglomerate rock from the summit, a piece of white stone embedded with pebbles of quartz and feldspar. This trip represented the completion of his tour of the high places of the then forty-eight states of the Union.

However, it did not represent the first time he had had to return to climb another mountain that proved to be higher than the previously accepted record holder. Fortunately not all such revisions have occurred in remote areas; several have taken place in long-settled states of the East. A few years ago it was discovered that the south ridge of Mt. Frissell, in the extreme northwestern corner of Connecticut, was the real high point in that state and not Bear Mountain, as had long been accepted. About the same time Jerimoth Hill, some twenty miles west of Providence, in Rhode Island, was shown to be a few feet higher than the supposed champion, nearby Durfee's Hill. The site of the Bok Tower, at Lake Wales, Florida, was recorded for decades as the highest land in Florida. Only recently, Britton Hill, between Tallahassee and Pensacola, has been shown to be higher. In each case Peters trekked back to climb the new elevation and bring home a substitute rock for his collection.

In the 1938 U.S. Geological Survey *List of Extreme and*

Mean Altitudes in the United States, Rib Mountain, near Wausau, was recorded as the high point of Wisconsin. Peters climbed to the shining quartz tip of its summit. Then, twelve years later, in the 1950 revised list, Sugarbush Hill was given as being higher. So Peters added 2,000 miles to his traveling distance and drove a round trip back to Wisconsin to climb the newly credited high point. For eight years this hill, on which we stood in the fading sunset, was the champion. Then, in the November 1958 revised list of the Geological Survey, Rib Mountain was again proclaimed the real high point of Wisconsin. Back in Freeport, George Peters restored to his collection his bit of Rib Mountain quartz and laid aside the smooth piece of glacier-transported reddish granite he had picked up in the open space among the sugar maples of this hilltop. Also on this hilltop he had climbed, as we climbed now, the 100-foot wooden Rat Lake Fire Tower. In all probability we were among the very last to ascend it. For close beside it a new steel tower was nearly completed, taking the place of the old structure, soon to be demolished.

We sat on one of the worn wooden steps up which Peters had climbed. Higher than the tops of the maples, we watched the sunset fade, watched swirling clouds, slaty-blue and bordered with flaming red, form and reform along the far horizon line. Across all the vast Northern Highland Province of the geologists, across 15,000 square miles of northern Wisconsin around us, the twilight deepened. Below in the valley small patches of mist formed like silvery spiderwebs that spread and joined and finally stretched in a single lake of vapor, glimmering faintly in the dusk. Above the mist, among the darker vegetation of the slope, fireflies gleamed. But here there were no rising sparks of insect fire, no shimmering dance of living light. Instead these luminous beetles winked from three to five times. Their lights waxed and waned like tiny coals breathed upon in rapid sequence.

The night breeze grew stronger. Still warmed by the land, it brought from a succession of wooded slopes to the south and west the songs of veery and wood thrush, hermit thrush and robin. Some voices were small and far away, others were strong and near at hand. Once, on some darkening lake, a single loon awoke the echoes with a long and lonely call. Well after nine o'clock, when the other thrushes had fallen silent, the veeries still sang on and on. Their liquid voices were all around us in the dark as we worked our way slowly down the wooden steps and descended what was then considered the highest of the Wisconsin hills.

TWELVE

WINGS IN THE SUN

IN the long northern twilight of another evening, 130 miles north and east of that Wisconsin hill, we wandered over the sand and pebbles and driftwood of the topmost shore of Lake Michigan. Three hundred miles straight to the south lay the lower shore, fringed by the Indiana dunes. We had virtually completed that ring around a lake that I had imagined so many years before. It was now mid-July. We were east of Manistique on the Upper Peninsula of Michigan.

We had traveled north from Sugarbush Hill past a Wild Rose Point, into a Whippoorwill Valley, through vacationland where cottage owners indicated they catered to overnight and week-at-a-time guests by displaying signs: "Sleepers and Weekers." We had crossed the Michigan line at Ironwood and at Watersmeet we had mounted that slight rise in the land from which three different watersheds slant away, one to Lake Michigan, another to the Mississippi River and a third to Lake Superior. We had spanned the Escanaba, famed river of Paul Bunyan's Inverted Forest, and we had watched more than a score of springs tossing the sand at the bottom of Kitchitikipi, The Big Spring of the Ojibways.

Now we walked along the shore in calm air after days of wind. Over the silky surface of the water slow ripples rode languidly in to die with a scarcely audible lisp at the edge of the wet sand. Into the placid air, all across the surface of the lake, tiny mayflies were rising and drifting like thistledown toward the shore. Pale yellow, frail and delicate, hardly half an inch in length, they were miniature relatives of the larger, darker mayflies of Kelleys Island. They alighted on our hands, our clothing, our fieldglasses, anchoring themselves for a final transformation. One clung to my finger as to a tree branch and I watched it slowly pull itself out of its former shell, leaving behind a white ghostly form, perfect even to the three slender threads of its trailing tail. In less than two minutes after its emergence it was on the wing.

When we swept the lake with our glasses we saw acres, square miles of mayflies, not densely concentrated but scattered everywhere. Each piece of driftwood on the beach was furry with the anchored husks of life. We sat in the sand beside one fragment of a board about three feet long and nine inches wide. It had been polished by water and silvered by sun and wind. I began counting the mayfly skins clinging to its surface. The total was 868, nearly 400 to the square foot. Speckling a projecting bit of rock, six inches by nine inches, there were more than 100 of these ghost mayflies. And when we came upon a dead herring gull, half buried in the sand, we discovered that all its visible plumage was overlaid with a multitude of the minute, translucent skins left behind by the molting ephemerae.

Among the night-blooming campion and the goose tansy scattered along the upper beach, we found harebells and rough cinquefoil and in the latter plants, anchored for the night, sound asleep, black-and-yellow flower flies. Each insect, at first glance suggesting a honeybee, had assumed the same position, facing in along the leaf, head close to the stem, six clawed feet anchored solidly to the sides of the leaf trough. At one place six of the banded flies were sleeping on a cinquefoil hardly two feet high. They were insect botanists, insect specialists. All along the shore these flies, and these flies alone, had turned to this one plant, this one particular kind of cinquefoil in choosing their lodgings for the night.

WINGS IN THE SUN

During this twilight walk we were nourished by that famed dish of the Upper Peninsula, Cornish pasty—pronounced *pass-tee*. A century or so ago, miners who emigrated from Cornwall, in England, to dig copper and iron in northern Michigan, introduced this "meal you can carry in your pocket." Shaped like a large apple turnover, it is filled with meat, potato and onion. No dish could be more satisfying or sustaining for active outdoor life. From that night forward as long as we were in the Upper Peninsula, this meal of the Cornish miners remained our favorite food.

Riding inland next morning we came to a community called Germfask. If you hunt over all the maps of all the lands, the chances are you will find no other place so named. The word is derived not from the Indians, not from a foreign land; it was manufactured on the spot, concocted from the first letters of the last names of the eight original inhabitants. To the west of Germfask, maps of Michigan show an area that is blank except for the fine, wandering lines of half a dozen southward-flowing streams. This is the great Manistique Swamp. Here, spreading across some 96,000 acres of sandy knolls and shallow pools, is the Seney National Wildlife Refuge.

It was eight-thirty that morning when we turned down the road leading to the sanctuary headquarters. Cordia J. Henry, the refuge manager, friendly, helpful, intelligent, seemed to us to represent the highest class of the professional conservationist in the Fish and Wildlife Service. For half an hour we watched the dash and scurry of a least chipmunk, red squirrels, and a silvery strain of eastern chipmunks outside his front door. Then, with Henry as our guide, we set out. All that morning he led us over the roads that wind for thirty-five miles among the twenty pools that impound nearly 6,000 acres of water in the sanctuary. Clear and richly brown, the color of strong tea, this water is stained by the roots and bark of the hemlock trees, by the leachings of the swamp.

On small islands, dotting the water, Canada geese were nesting. Starting with a flock of pinioned, captive-bred geese in 1936, the Seney sanctuary now has a population of nearly 3,000. The dramatic aspect of this achievement is the fact that no wild Canada goose had ever been known to nest in the area before. Because these wary birds prefer grassy islands with their view unobstructed, Henry was removing superfluous bushes and trees as a part of his game-management program.

Once we stopped to examine, among the weeds of the roadside, tiny flowers, each crowned with what appeared to be two small rabbit ears. These were the blooms of one of the figworts, *Scrophularia leporella*, a plant with a charming common name: Bunny-in-the-Grass. Nearby was a shadbush or Juneberry tree, its branches ripped and broken by impatient bears feeding on the fruit. Few animals are so destructive. A lumberjack we talked to in the north woods recalled finding an abandoned bunkhouse that a bear had turned inside out in a search for food. It had smashed a hole through one of the tar-papered walls in entering. When it left, instead of going out the same way, it had ripped another opening through the opposite wall.

High in the top of a red pine tree, at the edge of one of the pools, a huge mass of sticks formed the nest of a pair of bald eagles. Cradled on this lofty brushpile, two eaglets were already well grown. The fact that only here, in all our 19,000 miles of travel with the summer, did we see an occupied eagle's nest underscored the steady and tragic decline in the population of America's national bird. Within sight of this nest another pine tree, bare and dead, rose starkly against the sky. It, too, had held an eagle's nest. During a violent storm, about this same time of summer on a previous year, a thunderbolt exploded down the length of the trunk. It killed instantly the occupants of the nest, an eaglet and an adult, and their bodies were devoured by flames that swept through the mass of sticks.

The recorded instances of birds being struck by lightning are comparatively few. In the East, some years ago, two ornithologists on a field trip saw lightning strike an oak in which a hawk was perching. After the storm they found the bird lying dead at the foot of the tree. Near La Junta, Colorado, hundreds of sparrows and finches were reported killed when a bolt struck and ran along a wire on which they were perching. Charles L. Broley, the famed Eagle Man we had met on the Kissimmee Prairie of Florida during our trip with the spring, recalled one instance near Sarasota in which lightning struck a tall pine tree, killing a female eagle brooding on a nest. Another instance of one of these birds being killed by a thunderbolt is cited by Francis Hobart Herrick in his book, *The American Eagle*. On November 5, 1921, a deer hunter, Adrian P. Whiting, of Plymouth, Massachusetts, came upon a large white pine that had been blasted by lightning. As he was examining it he discovered the body of a bald eagle lying among the splinters of the shattered tree.

That birds have, on rare occasions, been stunned or killed by lightning in the sky is attested to by three records published in *The Auk*, the journal of the American Ornithologists' Union. About two o'clock in the afternoon, on April 11, 1941, Alexander Sprunt, Jr., reported, four men were inspecting a field of cabbages in the low country of South Carolina. Looking up as a sudden electrical storm broke, they saw a flock of large birds flying overhead. There was a flash, an almost instantaneous clap of thunder, and four of the birds, double-crested cormorants, fell headlong from the sky.

In two other instances, in widely separated locations, white pelicans were the victims. Near Nelson, Nebraska, on October 29, 1939, a storm struck the farm of Emil Schlief just as a flock of about seventy-five pelicans was passing overhead. One bolt of lightning struck the ground only about a hundred yards from Schlief's fourteen-year-old son, and at the same moment thirty-four of the great white birds began falling to earth.

Examination of the dead pelicans showed that, in some instances, the feathers were singed by the lightning. Ten years before and nearly 500 miles to the west, on August 16, 1929, near the Great Salt Lake in Utah, a filling-station attendant had just finished counting twenty-seven white pelicans passing at about 500 feet when there was a stab of lightning and a roar of thunder. He saw the whole flock come tumbling from the sky. Their dead bodies were found scattered over an area of nearly ten acres.

The shyest of all the birds inhabiting the great Manistique Swamp, one we had never observed, a bird we greatly hoped to see, is the little yellow rail. Smaller than a bluebird, it dwells in wet, marshy tangles, escaping on foot when disturbed, slipping away through the grasses, rarely flying, rarely seen. So elusive is it that even today much of the activity of its life is still unknown.

Henry pulled up beside one stretch of grassy lowland. He selected two white stones, each about half the size of a balled fist. With sharp clicks, suggesting the sound made by a child's metal "cricket snapper," he tapped the stones together: "Click-click—click, click, click—click-click—click, click, click." We were hearing a close approximation of the call of the yellow rail.

"Two weeks ago," he told us, "I could start them calling every time."

Now, just past the breeding season, the birds were silent. Although Henry continued his tapping sequence twenty or thirty times, although Nellie and I returned again and again that day and the next and repeated the performance, no sound came from the hidden birds of the swamp. We brought the two white stones home with us. They lie on the desk beside me now. They represent what is still our nearest approach to seeing or hearing a yellow rail.

By that noon we felt we could find our way about the convoluted maze of the ridges and pools. With a key to the entrance gate, so we could stay late and come early, we spent that afternoon and evening and the next day roaming alone among the deer and ducks, the loons and geese, the mink and muskrats and otter of this northern swampland. Aquatic beetles spun in the sunshine, spangling the brown surface of the water with glints and streaks of light. Once we saw a muskrat, carrying green plants in its mouth, swim for a hundred feet or more through this coleopterous fireworks.

All across the tens of thousands of acres of brown water and green swampland around us avian wings were also shining in the sun—the wings of birds soaring, birds turning, birds swooping, birds spiraling, birds alighting, birds lifting out of the grass and mounting over the treetops. They were the wings of kingfishers and nighthawks, of great blue herons and American bitterns, of snipe and marsh hawks, of greater yellowlegs and bluebirds and loons. We saw wings at rest in the sun, those of the floating Canada geese, the common terns, the hooded and common mergansers. Behind one of the latter trailed a long flotilla of twenty ducklings. These wings in the sunshine, wings at rest and wings in motion, form one of our dominant memories of Seney.

There were, we knew, other wings we rarely saw, those of the male black and mallard and ring-necked ducks, now in their eclipse plumage. We caught glimpses of them disappearing into cattail stands or heard their subdued quacking as we went by. Only among ducks is such a plumage known. Most birds lose their flight feathers gradually, one or two pairs at a time, but the male ducks shed their strong pinions all at the same time. For a period of nearly a month, while new feathers are growing in, they are unable to get into the air. The bright plumage of their breeding dress is replaced by less conspicuous, more somber garb that resembles closely that of the females. Thus, during an awkward, flightless time, the drakes are less conspicuous. But as soon as the wing feathers have been replaced another molt of body plumage brings back all the brilliant coloring of the winter and the breeding season. By shedding all their flight feathers at once, in this manner, these waterfowl begin their fall migration with new and perfect pinions.

Late that afternoon we sat beside a pond to watch a loon floating on the still water 100 or 150 yards away. Remembering the age-old stratagem known as the tolling of waterfowl, we tried to draw the loon toward us. Nellie thrust her hand outside the car window, waved a white handkerchief in the sun, then drew her arm inside again. The loon turned toward us. At intervals of half a minute or so, Nellie repeated the quick fluttering of the handkerchief. Pulled by curiosity, the loon swam slowly in our direction. For nearly a quarter of an hour we played this ancient game of the gunners, the bird moving closer and closer, sometimes head-on, sometimes quartering away, gradually lessening the distance between us. Before we drove on, it floated alert, ready for an instant dive, hardly more than fifty feet from shore.

We passed that same stretch of water again about eight o'clock that evening. The loon was there, this time swimming beside its mate. As we watched, a second loon passed by a hundred feet in the air, turned and circled the lake, calling as it flew. The bird below answered. Like echoes arising from the lake, descending from the sky, call followed call, loud and lonely sounds in the quiet of the summer evening. After the second loon was gone the paired birds rushed across the water in an almost upright position, sitting on their tails, then, close together, dipped and bobbed their heads in a nuptial display. It was a moving spectacle, these loons, the oldest in line of ancestry of all living birds, engaging in a display that extends back through millions of years, accompanied by the same wild cries that men of the Stone Age heard in the dusk of evenings infinitely remote.

We heard loons again in the mist of the next dawn. Stealing slowly through the watery shine of the vapor, we came upon deer, pausing with their mouths full of green leaves, heads uplifted, watching us approach. Snipe shot up, and Canada geese, protecting their goslings on either side, swam away into the curtaining mist. Once an otter crossed the road like an immense black measuring worm, it back arching high with every leap. As the sun rose we observed everything in a pale yellow light—the five flickers that left their ant-hunting in the roadway, the baby squirrel as taut as a steel spring, the thousands upon thousands of lowland spider webs, decorating, like spun-silver ornaments, every tree and stub and grass clump. The ground fog sank lower and the trees took shape around us, materializing slowly from top to bottom, reversing the normal procedure of growth.

In the mist and the dawn we heard another primitive voice —the faraway bugling of sandhill cranes. Closer at hand, late that afternoon, we heard these same birds when we followed a trail along the edge of an expanse of open wetland starred with the snowy tufts of cottongrass. It was this trail that led us to the adventure of the crane fields beside a remote pond near the southern boundary of the refuge.

We advanced at leisure, pausing now and then to pick handfuls of the blueberries that hung ripe beside the path. Eating as we went, we were dining on a favorite food of the cranes. Blueberries and grasshoppers make up the bulk of their diet at this season of the year. Although they are omnivorous, eating beetles, ants, dragonfly nymphs, caterpillars, grain, mice, crayfish—even, on occasion, small birds—the preponderance of their meals consists of vegetable food.

We followed the trail as it dipped into woodland and then, after a mile or so, rose to the top of a ridge of sand. Spreading away below us in the quiet of the sunset, we saw a wide and marshy pond. Along its farther edge four deer and half a dozen cranes stopped their feeding and lifted their heads to watch us. A shrill, loud, far-carrying "Gar—oo—oo—oo-a-a!" had sounded the alarm. This call of the sandhill crane has been most often compared to that of a bugle or trumpet.

We stretched out and lay motionless in the warm sand. Soon we seemed forgotten. The deer lowered their heads and the cranes stalked about in their food hunting, stabbing or probing the ground in their search. The scene around us was such as the pioneers had known. The voices of the birds now came across the water in a low, guttural muttering, a kind of purring "per-r-r-ump! Per-r-r-ump!" that brought to mind the distant, rhythmic sound of flying brant. This is the call of the cranes when all is well.

As the sun sank, the gray birds, appearing almost as tall as a man, stalked with stately, slightly jerking tread beyond the yellow spikes of the loosestrife, the swamp-candles, of the pond edge. Their heads were thrust quickly forward with every step. They moved deliberately. In emergencies, however, a sandhill crane can advance swiftly. It can even outrun a man. And among birds it is relatively long-lived. One individual in the Washington, D.C., Zoo reached the age of twenty-four years, two months and eighteen days and another, at the New York Zoo, died at the age of sixteen years, one month and one day.

At short intervals the feeding birds paused and lifted high their red-capped heads for a wary survey of their surroundings. One, dissatisfied with its feeding ground, flapped into the air, traveled a few hundred feet and dropped to earth again directly between two browsing deer. Another took off with a leap into the air and a powerful down-drive of its great wings. Rising higher and higher, it curved in a wide circle toward us. Unmoving, we watched the huge bird, its neck and feet far outstretched, turn above us. Its wings, having a spread of more than six and a half feet, moved steadily with quick, flipping upstrokes and powerful downbeats. Cruising in unhurried flight beside a highway, one crane was timed at thirty-five miles an hour. As the bird above us swung away, wheeling upward, its wings were caught and turned rosy by rays from the west. They were the largest and the last of the wings we encountered that day in the Seney sun.

Driving out to Germfask over dark roads that night we paused where a stretch of woodland echoed with the song of the whippoorwill. Farther south the singing season of this bird had ended. Here in the north woods, although near the close of the breeding period, the males still were calling. For a quarter of an hour we sat in the deep shadows of the road listening. Only a few times more that summer would we hear this voice of the dark. One of the songs of the Iroquois Indians—part of a ritual of fire and darkness—was "The Chant of the Whippoorwill." Each stanza ended with the same words: "Thus sings the whippoorwill—Follow me! Follow me!" There is, in truth, a bewitchment that many feel when they listen long to the repeated incantations of this hidden singer. Of all the things that come to life in darkness, what three better represent the magic of the night than the gleam of the firefly, the light of the will-o'-the-wisp and the echoing cry of the invisible whippoorwill?

More than once at the end of days like these we thought of Richard Jefferies and that masterpiece of English nature writing, *The Pageant of Summer,* set down during the last months of his life. Of sun-filled days in the open he wrote: "Never could I have enough; never stay long enough. The hours when the mind is absorbed by beauty are the only hours when we really live, so that the longer we can stay among these things so much the more have we snatched from inevitable Time." All these days of our summer wanderings seemed to us pure gain, days that we, too, were stealing from inevitable Time.

THIRTEEN

MYSTERIOUS MAPLES

"THE horror of that moment," the King went on, "I shall never, *never* forget!"

"You shall, though," the Queen said, "if you don't make a memorandum of it."

Not infrequently during our long trip with the summer, these words of the White Queen, coming down the years from behind Alice's wonderland looking glass, recurred to my mind. I remembered them now as I slipped a pencil stub and a small spiral-ring notebook back into my pocket. The English poet, Thomas Gray, was right: For anyone interested in accurate observation, one note set down on the spot is worth a whole cartload of later reminiscences.

At that moment we were standing in the hush of the great north woods of the Upper Peninsula. Close by our trail a path of claw marks ascended the silvery-barked trunk of a beech tree. It recorded a black bear's ascent in search of nuts. But it was not this tree that held our absorbed attention. Instead it was a nearby maple. At first glance it looked like all the other tens of thousands of sugar maples we had seen in recent days. But, as we drew close, Cordia Henry pointed out a small patch where the bark had been scuffed away. There, scattered thickly over the bare wood, were tiny bumps. The tree, the first of its kind we had ever seen, was the rare and valuable bird's-eye maple.

If you draw a circle with a 200-mile radius, using Escanaba

MYSTERIOUS MAPLES

on the Upper Peninsula as the center, you will enclose the source of more than 90 per cent of all the bird's-eye maple in the world. It was from Escanaba, in 1936, that 100,000 square feet of this wood went to the Cunard Line in England to finish the cabins of the S.S. *Queen Mary*. All over the north country we encountered references to this maple. But only here did we see the bird's-eye wood in a living tree.

Michigan, among the fifty states, ranks fourth in the production of maple syrup. Why it so far exceeds all others in the production of bird's-eye maple wood, why the world's greatest concentration of such trees is found growing in such a relatively small area on the continent, are still unanswered riddles. The whole story of this sought-after wood is filled with mysteries.

For the bird's-eye maple is not a distinct species. It is not a strain of the sugar maple. Its characteristic—wood with the grain contorted and mottled with little spots or "eyes"—is not inherited, not passed on from parent tree to seed to sapling. Some years ago when one experiment station on the Upper Peninsula planted seeds from bird's-eye trees, they obtained nothing but ordinary sugar maples. Moreover, the trees never appear in groves or stands. They are scattered and individual. Around them there are always other maples with no bird's-eye characteristics at all. What makes these particular trees grow in this particular manner?

Since that morning on the forest trail, I have asked that question of many persons. I have asked it of botanists. I have asked it of lumbermen. I have asked it of research scientists. I have asked it at the world's largest processing plant for such wood, The Bird's Eye Veneer Company, of Escanaba. Everywhere the answers I received only deepened the riddle, only underscored the strangeness of the story of these mystifying trees.

On the outskirts of Escanaba, the hub of the bird's-eye maple world, the veneer plant was shutting down for the noon hour when I talked with Russell Lee, the man in charge, and Clint Paulson, the company's veteran timber cruiser who has roamed the Upper Peninsula for years in search of the rare trees. Later Paulson took me through the yard where the logs were coming in from the forest. In his sixties, square-shouldered, a bit stooped, big-handed, soft-voiced, he had the appearance of being made of iron. He had spent all his life in the north woods.

Usually, he told me, he discovers the greatest number of bird's-eye trees in rolling land where hardwoods are mixed. The characteristic is found never among soft maples, always among hard maples, sugar maples. But wherever hard maples grow, bird's-eye trees are possible. Although he has never encountered them in stands or groups, he thinks that where he finds one he has a better chance of finding another. At such a place he always inspects the surrounding trees with special care. North slopes appear to produce the most bird's-eyes. But Paulson has found such trees rising from ground that slanted downward toward all points of the compass. So the direction of the slope cannot be the deciding factor.

Could it be the kind or condition of the soil? Apparently not. Some of the bird's-eyes he has discovered have been rooted in clay soil, some have been anchored in sand, some have derived their nourishment from black loam. Gravel has been a main feature of the ground in some areas while it has been entirely absent in others. Virtually every type of soil known to the Upper Peninsula has, in one place or another, nourished bird's-eye maple trees.

As we talked, we walked about the yard among great piles of logs. Printed on the ends of many of them were the letters "B E"—signifying bird's-eye maple. Most of these logs had come from the other side of the peninsula. On Grand Island, in Lake Superior near Munising, virgin timber was being felled. Fully 90 per cent of the Michigan bird's-eye, I was told, comes from virgin stands of maples. There is relatively little bird's-eye in second-growth woods. In the course of time, Russell Lee speculated, bird's-eye maples may become largely a thing of the past.

Not only do seeds from such trees fail to produce similar trees but saplings that show bird's-eye characteristics usually develop into normal maples when they are transplanted. A generation ago, one resident of upper Michigan conceived the idea of providing a fortune for his children by planting a grove of valuable bird's-eye maple trees. He marked saplings, had workmen dig circular trenches around them in the fall and then, when the ground froze, had the small trees, with their roots in a ball of earth, transplanted onto his land. The saplings prospered but the idea did not. As they grew, virtually all of the trees lost their bird's-eye characteristics and changed into ordinary sugar maples.

In some instances a maple will grow for half a century, producing the bird's-eye formations all during that time, then change and continue to grow as a conventional tree producing conventional wood. I talked to one lumberman who had brought home, with considerable difficulty, a large log from the forest and then discovered that the side he had blazed contained bird's-eyes, as he had seen, but that the other side contained none. Sometimes the characteristic extends not only through the entire trunk but into the branches as well. In other cases the bird's-eye wood occurs in patches in the trunk, the patches separated by normal wood. The size and number of the eyes also vary greatly in different trees.

In the Seney region, a few years ago, a farmer burned bird's-eye maple in his kitchen stove most of the winter. He had felled a large tree and sawed it into short lengths for splitting into cordwood before he discovered its value. On the outside a bird's-eye and an ordinary maple look the same. Only by blazing the tree, removing a section of the bark, can the character of the wood underneath be established.

During all his years of timber cruising Paulson has observed only two things that might be called clews to the bird's-eye. One he calls "the apple tree"; the other, "the hobble skirt." He has noticed that many—but not all—of the bird's-eye maples have a noticeably smaller crown. In the distance such a tree, as he puts it, seems to have an apple tree growing at the top of the maple. This peculiarity is the one he looks for first. Where it exists it can be noticed as far as the tree can be seen. The second clew is not apparent except in close examination. The lower trunk of a number—but again not all—of the bird's-eyes have a kind of "hobble skirt" appearance. The trunk rises at a uniform diameter for two or three feet; then there is an abrupt variation in the diameter—as much as two inches. Wherever he has found such a tree trunk he has discovered, on removing a patch of the bark, that the wood was characterized by bird's-eyes. It is also his general impression that the sought-after trees are likely to have slightly smaller leaves than other hard maples.

Among the sugar maples of the Upper Peninsula of Michigan it is estimated that as many as 20 per cent may have some bird's-eye characteristics. However, Russell Lee told me, not more than one-half of one per cent of them are suitable for making veneer. The eyes are too few and too scattered. The finest of all bird's-eye, with the greatest concentration of eyes, is known to the trade as picture wood or clock wood. Originally it was used in making fancy clocks and elaborate picture frames. In producing veneer from logs the layer of wood is cut away as the log revolves so the sheet is drawn off like paper from a roll. Each bird's-eye formation extends like a slender rod or pencil outward from the heartwood. The milling of the logs has to be done with special care. If the high-speed cutting tools slow down ever so slightly, they tend to pull out the bird's-eyes. Furthermore, the veneer must always be laid as it comes off the log. If it is turned over and then smoothed and sandpapered, the eyes are likely to pop out, leaving the sheet riddled with holes.

MYSTERIOUS MAPLES

As a rule, I was told, it is the lower part of the trunk that tends to have the most bird's-eyes. Sometimes only this portion of the tree can be used in making veneer. There have been cases in which the heartwood alone contained the bird's-eye formations. Russell Lee expressed the theory that bird's-eyes may be the result of "suppressed growth." Most of the trees are found in virgin stands where the forest is thick and young trees are deeply shaded. Originally, he believes, some of these saplings received insufficient light and for a time were unable to develop normally. Misdirected growth, producing rodlike bird's-eye formations, resulted. The trunks of such trees seem filled with small, abortive "branches" that remain contained within the wood itself.

Exactly what are these "buried branches"? Are they really branches? If not, what are they? To those questions varied answers are given. The mystery of what causes the bird's-eyes in some trees and not in others is balanced by the mystery of what it is that is produced. Later, at the New York Botanical Garden, the Brooklyn Botanic Garden and among the textbooks on timbers and woods and botany in general, I sought for a final answer. But none was forthcoming. The hypotheses were contradictory. The botanists were far from being in accord. Each small circlet on the face of the maple veneer represents an enigma.

For generations the theory has been advanced that latent buds within the trees are stimulated into misdirected growth, perhaps by a virus, perhaps by a fungus, perhaps by insect injury. Somewhat in the manner that a burl or witch's broom is produced, this abnormal activity creates from the latent buds the twiglike formations that extend outward through the successive layers of the developing wood. However this "adventitious bud" theory is flatly rejected by a number of modern authorities. The Dean of the Yale University School of Forestry, Dr. Samuel J. Record, writes in his textbook on *Timbers of the New World*: "The cause of bird's-eye is not definitely known, but it is not, as so frequently stated, attributable to buds." And so we are left with questions without answers, with the double mystery that is presented by every bird's-eye maple.

As I looked at the living tree in the Seney forest that day, I wished profoundly to know its inner history, masked by bark, hidden within its roots and trunk. Silent, so far removed from our thinking minds, withdrawn into itself, its life so different, it held, without an outward hint, the secret of this twofold enigma.

The same mood returned on a night when we slept at a north-country motel where all the walls of our room were covered with bird's-eye maple. In the late-afternoon sunshine, I had stood with the owner on the little walk that ran behind the building. As we talked, the man's right foot was rarely still. With the toe of his shoe, he reached to the right to step on an ant, reached to the rear to crush a beetle, ran up and down the edge of a crack to annihilate a spider. For one whole winter, the man with the destructive toe told me, he had hunted in the woods to find enough bird's-eye maple to finish the rooms of his motel. He had retired, not long before, from the lumber business. He recalled the great boom in bird's-eye maple in the mid-twenties when the demand for ornate radio cabinets was at its height. Now, competing with plastics and synthetic veneers, the wood of the mysterious maples is somewhat less frequently employed. But it is still eagerly sought all across this northern land where, for some inexplicable reason, it is most abundant.

For a long time that night we looked at the beauty of the walls around us. Scattered or clustered, the thousands of enigmatical little eyes caught and held the glow of the shaded lamp. Then, at last, we switched off the lights. And so we slept surrounded by beauty and by mystery. And when we awoke we saw in the dawn light the beauty and the mystery still around us.

FOURTEEN

THE DARK RIVER

THE Great Lakes descend like giant steppingstones toward the east. Lake Superior is the top step. The waters of Lake Michigan and joined Lake Huron are twenty-one feet lower. Lake Erie is eight feet lower still, while the drop from Erie to Lake Ontario, at Niagara Falls and the gorge of the Niagara River, represents the giant step, a descent of more than 320 feet.

Ever since we paralleled the St. Lawrence and came to the far eastern end of Lake Ontario, we had been climbing the stairs of the Great Lakes. We had worked west along the southern shores of Ontario and Erie, mounted northward beside the western edge of Lake Huron, made a clockwise circle of Lake Michigan and now, at the northern fringe of the Upper Peninsula, we stood on the shore of Lake Superior. Our first glimpse of this last and largest of the five Great Lakes came as we lifted over the brow of a hill on a forest-walled gravel road running north to Grand Marais and its nine miles of dunes, the "great sand," the Gitche Ganow of the Ojibway.

To fill the beds of the five Great Lakes, someone has calculated, would require all the flow of all the rivers of the world for an entire year. The largest flood of all would be needed to replenish Lake Superior, with its area of 31,800 square miles—larger than that of South Dakota—its shoreline of 1,500 miles and its extreme depth of 1,008 feet. It alone contains enough water to sustain Niagara Falls at its present rate for almost 100 years. The biggest, deepest and coldest of the Great Lakes, it is the largest body of fresh water on the globe. It exceeds by about 5,000 square miles its nearest rival, Lake Victoria in Equatorial Africa.

Along the Superior shore, back to the Lake Michigan shore, around and across the Upper Peninsula, down its edges and from side to side we traveled during the succeeding days. For us this area extending west from the top of Lake Huron, above Lake Michigan and along the Lake Superior shore was all new country, wonderfully new, wonderfully wild. We were reluctant even to glance down at the map while riding lest we miss something of interest. Everywhere we felt far away from cities and twentieth-century civilization. And everywhere we were surprised by distances, by the bigness of this upper portion of Michigan. In length, it is greater than the north-and-south span of the rest of the state down to the Indiana line. In area, the Upper Peninsula is twice as big as the state of Massachusetts.

As we ranged back and forth over it, we encountered miles of lowland covered by leatherleaf as by a dense cultivated crop. We paused at little bays in the forest filled with the hum of bees and the pink of milkweed flower clusters. We crossed Teaspoon Creek and came to Laughing Whitefish River. We saw the great locks at Sault Ste. Marie and the gulls that swept low, like pigeons, over the city streets. With morning mist entangled in the treetops and crows flapping up from feasts where porcupines had been killed on the road, we traversed the Hiawatha National Forest.

All this Upper Peninsula is Hiawatha country. It was on Henry Rowe Schoolcraft's monumental work on the American Indians of this region that Longfellow based his poem. Everywhere we went we encountered the legendary settings of Hiawatha's adventures. The lake we first glimpsed here at Grand Marais was Gitche-Gumee, "the shining Big-Sea-

Water." The yellow dunes to the west were born when, at Hiawatha's wedding, the dancing feet of Paupukkeewis tossed sand aloft until it drifted over the landscape, "heaping all the shores with sand dunes." A score of miles farther west we came to other surroundings linked with the Hiawatha legend, the Pictured Rocks, the Grand Portal, the cliffs of Cambrian sandstone that lift in a sheer yellow-red wall for as much as 200 feet above the water.

From Munising we rode by boat one afternoon along the cliffs and caverns of this portion of the shore. Green, yellow, brown, white, blue, vertical bands of chemical deposits ran along the lower part of the cliffs. Sweeping in from the open lake, winds and waves have eroded the sandstone in many places into odd, towering formations—Battleship Rock, Pulpit Rock, Flower Vase Rock and Miner's Castle. At the water line, storm waves, eating into the high palisades, have formed caverns such as the U-shaped Indian Drum where the smashing water produces rolling peals of thunder that can be heard for miles. To the Ojibway or Chippewas, these caves were the legendary dwelling places of the gods of thunder and lightning.

The yawning doorway of the largest of all these caverns was early named the Grand Portal. Pierre Esprit Radisson, the Frenchman who explored this Great Lakes region between 1652 and 1684, was the first white man to see the Pictured Rocks and the Grand Portal. The arch of the entrance, he noted at the time, was so lofty that "a ship of 500 tons" could have sailed beneath it. In 1906, two and a half centuries after Radisson, the roof of the arch, weakened by long erosion, collapsed and blocked the entrance of the cavern with fallen rock. Today the ragged cliffs that tower up are famed for the gulls of Hiawatha, thousands of herring gulls that nest each year in the niches and on the ledges of the eroded rock.

We saw them brooding in little crannies far above us. We saw them soaring on the updrafts before the cliff face. We watched their shadows sliding along the wall of rock above the vertical stripes of many colors. We saw them dropping away from their nesting niches, plunging in long nose dives, first close to the rock, then curving out away over the water —water vivid emerald green below the palisades, clear blue under the open sky. We passed fledglings, gray-brown in their immature plumages, floating on the lake. Only a few days before they had begun to appear from the nests of the Grand Portal just as other broods had come from other nests along this portion of the Superior shore in years so long ago that the legend of Hiawatha was known only around the campfires of the Indians.

Great seas build up across the Superior waters when storms sweep in from the northwest. At such times the impact of the waves against the sandstone crags sends water exploding upward along the walls of rock for as much as 100 feet. Nestling gulls, during these gales of early summer, are sometimes swept from their crannies and killed by wind and water. Our day on the world's largest lake was a placid voyage. But Superior is the stormiest, the most uncertain of the five Great Lakes, and the gulls of Hiawatha often ride on booming winds and over tumbling water in the vicinity of these cliffs.

On our outward journey the birds paid scant attention to our boat. But on the return trip, about the time we passed the high, eroded turrets of Miner's Castle, they started winging toward us, converging on our wake. At this point, during each homeward trip, the captain tosses up fragments of bread. With swift swoops and sudden checks the birds all tried to snatch them from the air before they reached the water. The boat going east is of no interest to the gulls; the same boat going west, just when it reaches Miner's Castle, becomes the center of attention.

We came back to Miner's Castle later that day by land. Between its dusty battlements of weathered sandstone we peered down over an unprotected brink at the churning water a hundred feet below. As we stood there a thin, nervous man

FAWN of the white-tailed deer on the Door Peninsula of Wisconsin. Its age, when photographed, was about two weeks.

SUNSET LIGHT reflected on the water of a slow stream that winds in a shallow flow across a northern bog.

arrived with his wife and two small daughters. The little girls scrambled to the edge to look over. The man glimpsed the void a dozen feet ahead of him. He fell on his knees with a screech, crawled turtlewise back from the precipice and sat, with his family clustered around him, mopping his brow and saying:

"All of a sudden I felt scary!"

At last he was helped to his feet. The little group disappeared down the trail to the parking field. A few moments later we heard screams and a crash. The nervous man had shifted into the wrong gear. Instead of backing away from the precipice, his car had shot ahead, halted at the very edge by a barricade of posts and heavy wire mesh. For several minutes he sat with his head in his hands. Then slowly the car backed in a semicircle and slowly drove off down the road. With relief we watched it creep away, everybody safe and sound for the moment.

Our last memory of that day was not of the nervous man of Miner's Castle but of a small boy in paradise. As we ate supper in a Munising cafe, the owner's son, about eight, wandered about behind the ice cream counter. While the customers watched in fascination, he concocted for himself a supersundae. First he ladled in vanilla, chocolate and strawberry ice cream. Then he covered it with chocolate syrup, added pieces of banana, sprinkled it with ground nuts, heaped it high with whipped cream, tinted it with red coloring matter and finally garlanded it with a ring of maraschino cherries. He viewed his handiwork with satisfaction and took a first bite.

"I don't know what you call it," he replied to one of the waitresses, "but it's good!"

From the Pictured Rocks we wandered east again toward Whitefish Bay and the great Tahquamenon Swamp. In all the Hiawatha country of the Upper Peninsula, this region, only thirty-five miles from the Straits of Mackinac, remains the most primitive and untamed. It is still the haunt of the timber wolf. If there is a wolverine left in all the Wolverine State of Michigan—which is doubtful—it probably makes its home here. The thousand square miles of this tangled wilderness occupy a shallow bowl bounded by rolling hills of limestone on the south and by a tableland of sandstone on the north. Draining this bowl is the brown Tahquamenon, the Dark River or Golden River of Hiawatha. Longfellow spelled it Taquamenaw. It was on the banks of this wilderness stream that Hiawatha built his legendary canoe of birch bark. Today, endangered by the new bridge across the straits, the area may well become another shot-out "sportsman's paradise" unless steps are taken to preserve it as a wildlife sanctuary.

At the Seney refuge, Cordia J. Henry had told us the best way to see the Tahquamenon was to take a boat cruise to the falls. I had in my pocket a note he had given us to Ken Slater, whose revamped LST landing craft were making the four-and-a-half-hour run twice a day. With clear weather and heavy tourist traffic, these days, the boats were loaded on every trip. When we stopped at Hulbert to ask directions to the dock, the woman we inquired of volunteered:

"This year they are doing a boomerang business on the river. Yessiree, a boomerang business!"

Ten miles of gravel road carried us north to the river landing. Ken Slater, a sturdily-built former timberman and a keenly observant naturalist, was finishing loading up for the next trip. Through his winter hobby of recording on tape the reminiscences of the old-timers, he has become an authority on the pioneer history of the peninsula. With him we climbed aboard the steel-hulled, scow-shaped craft. Ponderously it backed out into the river. Slowly it turned with the brown current, gained momentum and headed away on the long serpentine of its cruise through uninhabited land.

All down the river miles that followed, a wilderness panorama unfolded on the banks beside us. Beaver meadows spread away, old ponds that, generations before, had become filled with silt. Once we noticed a dozen white cedar, or arbor vitae, trees that ran as though on a straight ruled line. They all had sprung from the long-decayed ridge of an ancient beaver dam. Several times, as we rounded turns on the winding stream, deer bounded away across openings or peered at the passing boat from behind a thin screen of foliage. Logging trails, not yet completely obliterated by time, ran down to the river's edge. They reminded us of a past era when the felled forests of this region floated to mills in great spring drives down the Tahquamenon River.

At intervals along the banks we came upon "witness trees," the blazes made by pioneer state surveyors more than a century before still apparent on their trunks. When the bark that has grown over these scars is pried off, its inner side reveals, in a perfect negative, the scratches of the code-mark put there so long ago. We thought of the adventures of those early surveyors, combing this wilderness and marking such witness trees at the corners of each square mile. It has always seemed to me a great loss that pioneer surveyors so rarely set down their experiences. The homesteaders, the Argonauts, the big-game hunters, the gold-seekers, the explorers, all these have left numerous and invaluable journals. But early surveyors, whose adventures must have equalled any of these, have been curiously silent. Hardly a one has left a written record of his travels in the wilderness.

Half a dozen times, where the bank shelved steeply down, we observed otter slides. Ken Slater pointed out one where he had seen a whole family, a mother and three pups, at play that morning. As the boat approached, the mother rushed about in chattering excitement trying to herd her frolicsome litter to safety. But the young otters were reluctant to give up their play. One pup would come shooting down, tobogganing headfirst into the brown river. She would get it headed safely up the shore when another would streak down the slippery bank. She was still struggling to round up all three of her romping pups when the boat went by. Not far from here, on another trip, Slater saw an old otter come to the surface with an unusually large fresh-water clam. The bivalve had its shell clamped tightly and Slater watched closely to see how the animal would pry it open. He saw the otter turn the shell around and around in its forepaws. Then it began biting at the rear. In a few moments it had chewed off the hinge and opened the shell.

Winding continually, as swamp streams are wont to do, the Tahquamenon occasionally cuts off one of its own loops in times of high water. The island that is created in this manner is enclosed, as within a moat, by the horseshoe curve of the former channel. We passed one place where the floodwaters of spring had opened a channel across three successive loops, giving rise to three islands of descending size—Big Island, Little Island and, the smallest of all, Penny Island.

Twice, before we reached the falls, we encountered the huge stick masses of old bald-eagle nests. But neither was occupied. In past decades, the snowy head and tail of our national bird was a common sight along the Tahquamenon. However—again reflecting the tragic decline of the species—bald eagles are now but rarely seen along the river and the nests of former years have been deserted.

The falls of the Tahquamenon, 200 feet wide, plunge over the brink of a forty-foot cliff of sandstone with a continuous thunder that fills all the surrounding forest. This wilderness cataract is second only to Niagara east of the Mississippi. For us, the thing we will remember longest about this famed scenic attraction of the north woods is the play of colors across the falling water, golden foam glinting in the sunshine, the torn currents arcing downward in delicate shadings of brown—that and a spotted sandpiper that flew up and down the stream over the lip of the falls so low it seemed drenched each time with spray.

That evening, with Ken Slater and his wife, Ada, we left their home—a house almost ringed by bird feeders—and followed the curving shore of Whitefish Bay toward the east. It was dusk when we reached our destination, Halfaday Creek and the bridge that spans it close to its mouth. A high, steel-mesh fence stretched across the stream and a red warning light glowed in the late twilight. On a padlocked gate we made out the words: "WARNING. DANGER. HIGH VOLTAGE. Electrical Sea Lamprey Control Device. U.S. Dept. of Interior. Fish and Wildlife Service."

We were standing on a silent battleline that extended away along the dark Superior shore. Other warning lights shone at the mouths of other streams—Betsy River, Hurricane River, Chocolay River, Big Garlic River, Two-Hearted River, Laughing Whitefish River. Each marked the testing ground of an experimental weapon being tried against an invasion from the sea, against a fearsome explosion of the sea lamprey population that had already virtually wiped out the multi-million-dollar fishing industry of the Great Lakes.

Resembling the eel in appearance, the lamprey, *Petromyzon marinus*, is one of the most ancient fishlike dwellers of the sea. It is so primitive that its skeleton is formed entirely of soft cartilage instead of solid bone. Its sucking mouth is devoid of jaws. Instead it forms a round opening equipped with concentric rings of pointed teeth. With this mouth clamped to the side of a fish, the bloodthirsty lamprey employs its file-like tongue to rasp a hole into its victim's body through which it drains away vital fluids until it is satiated or its prey is dead.

Originally this nightmare creature out of the remote past left the ocean and entered coastal streams only to spawn. Then a few began spending their entire life cycle in fresh water. A century and a quarter ago, the digging of the Welland Canal around Niagara Falls opened the way into the upper Great Lakes for the lamprey eels. Because warm and shallow

Lake Erie provides an inferior environment, they were nearly a century working west to the deeper, colder waters of Lake Huron. There, under favorable conditions and with most of their natural parasites apparently left behind, the explosive multiplication of the lampreys began. They spread to other lakes. The disastrous consequences soon became apparent. In two decades the annual catch of lake trout in Huron waters dropped from 2,000,000 pounds to less than fifty. In the 1930's, commercial boats in Lake Michigan brought in as much as 6,000,000 pounds of fish; in the 1950's the annual catch sank to less than 100 pounds. A whole industry collapsed as the spreading blight of the sea lamprey upset the balance of nature in the western Great Lakes.

Fighting to stem this tide of destruction, government scientists studied every phase of the sea lamprey's life cycle. They watched it choose gravel riffles in swift streams for spawning. They observed it hollow out an oval nest by attaching its sucker mouth to stones and in this manner moving them downstream. They watched the male and female lampreys clamp their suction discs to the same larger stone and, side by side, eject the fertilizing sperm and as many as 100,000 eggs into the nest. The blind baby lampreys hatching from these eggs drift downstream, burrow into sand or mud banks and spend the next four or five years there as they grow slowly to a length of from six to eight inches. Then, under cover of the winter ice, they slip down to the lake to live the life of predators for a year and a half before they return upstream to spawn and die. Because the adults hunt singly and never travel in schools they are particularly difficult to destroy in the lakes. It is in the streams that they are most vulnerable. This is the weak link in their life cycle that the scientists have attacked.

In 1944, close to the mouth of one Michigan stream, they constructed a barrier trap. During the first spawning season 3,366 lampreys were caught and killed. Five years later, a lamprey-tight weir in a river took 29,643 adults that were ascending the stream. But such weirs proved costly to install and maintain. Moreover, they were not practical for all streams in which lampreys spawn. Innumerable other suggestions were considered—the introduction of lamprey parasites, the use of flame throwers to destroy the teeming inhabitants of the mud banks, the offering of a bounty for all lampreys killed. But it was to electricity that the government scientists turned next in their search for a more efficient weapon.

The primitive lamprey is less sensitive to mild electric currents than are the more highly organized trout and other gamefish. The plan being tried in the streams of the Upper Peninsula was to draw across them an invisible barrier of electric current. The valuable fish would be warned away at the fringes while the more insensible lampreys would enter the lethal fields and be stunned or killed. The electrical weir beside which we stood at Halfaday Creek was one of more than fifty similar devices being tested on the peninsula, each that night standing silently on guard under the red glow of its warning light.

The low position of the lamprey in the evolutionary scale in the end proved its undoing. Attacking it from this angle, research workers concentrated on developing selective poisons, chemicals that would kill the young lampreys without harming gamefish. One winter day, after our trip was ended, I sat in the Washington office of an official of the U.S. Fish and Wildlife Service and heard the story of the seemingly impossible search for such a poison and its successful conclusion. Month after month and year after year chemical compounds were tested. A thousand, five thousand, more than six thousand different combinations were tried before the right formula was discovered. In laboratory tests it destroyed every lamprey in forty-five minutes yet left gamefish unharmed after an exposure of twenty-four hours.

It was during the week of October, 1957, that this selectively toxic chemical received its first field test in a Michigan stream. As little as thirty parts of the chemical added to 1,000,000 parts water in the creek were sufficient to bring death to virtually every lamprey larva hiding in the sand and mud banks. In all sizes, their dead bodies drifted downstream and collected in the eddies. The test was, as the government report puts it, "unconditionally successful." With the destruction of the larvae, the menace of the sea lamprey can be eliminated and the balance of nature restored. But the comeback of the Great Lakes fish will, in all probability, be a long, slow recovery. Even with restocking, it is expected to take twenty years.

Once as we stood beside another bridge and another glowing light of red, a great horned owl loomed suddenly above us, immense, crossing over the road on silent wings. One moment it was silhouetted against the sky, the next it was lost beyond the black treetops. From far away the voice of a whippoorwill reached us and in the darkness of the forest the barred owls began their calling. Closer at hand, whenever we stopped, there was another voice, thin and steely and multiplied a hundredfold, the voice of the mosquito.

In pioneer times, the Upper Peninsula of Michigan was celebrated for its clouds of such ravenous tormentors. Radisson wrote eloquently of the misery produced by the "maringoines," as he called the mosquitoes in his *Voyages of Pierre Esprit Radisson*. Not far from the present site of Sault Ste. Marie, he tells us, he came upon two men who, for want of covers, had "buried themselves by the watter side to keep their bodyes from the flyes called maringoines, which otherwise had killed them with their stings." On this night, we listened to the shrill chorus of the "maringoines" around us from behind the protection of a chemical repellant frequently rubbed on our hands and necks and faces.

In *The Naturalist in Nicaragua*, Thomas Belt recalls a conversation with a native helper who had spent a troubled night slapping mosquitoes.

"Mr. Belt, Sir," he asked, "can you tell me what is the use of mosquitoes?"

"To enjoy themselves and be happy."

"Ah, Sir, if I was only a mosquito!"

In the course of enjoying themselves and being happy, the mosquitoes of the world exhibit remarkable capacities. They can detect warmth at a considerable distance. Through darkness they can follow by scent a trail of breathed-out carbon-dioxide. The stiletto proboscis that pierces the skin, causes pain, raises lumps and in some cases produces illness, weighs a mere 1/6,000,000 of an ounce. The four rapier-like mandibles it contains can penetrate through human skin in two seconds. To produce healthy eggs the female mosquito requires a diet of blood. If given a choice she prefers the life fluid of birds to that of humans. Different kinds of mosquitoes, it has been discovered, tend to feed at different times during the night. Because the wigglers that hatch from the eggs must thrust their breathing tubes through the surface film, they can live only where the water has little or no current. Here on the Upper Peninsula, mosquitoes have no doubt owed a great deal to the beavers. Their ponds of impounded water provide almost perfect breeding places.

By the time we had worked back to the town of Hulbert in a roundabout return, it was late at night. Once again, as so often on these wandering trips of ours, we had encountered congenial people in the Slaters. Our lives touched at this one point in their long continuity. We might never meet again. Yet how often in memory would these hours return! My last recollection of that day was listening to a high-fidelity recording of a symphony orchestra while we ate wild strawberry sundaes at midnight.

FIFTEEN

BEARS OF COPPER HARBOR

THE happy men—who are they? A hundred times during our travels with the seasons I asked these same questions of those with whom we became acquainted along the way:

"Who is the happiest person you know? What does he do? What is he like?"

I scarcely expected to find a common denominator of happiness, and I did not. Those who were named were infinitely various. Many were poor, a few were rich, several were sick, one or two were not what would be called strictly law-abiding. With only a solitary exception, in which one person named himself, the happiest person was always someone else. The only similarity I could find was that almost all of those mentioned had lost themselves in activities which they enjoyed tremendously. These activities were often of no great importance. Not infrequently they appeared trivial and even ridiculous to others.

As evening fell on this day at the very tip of the Keweenaw Peninsula, the northernmost arm of Michigan, we met a man who, for the time being, seemed to qualify for this select group. His was an unexpected and bizarre absorption. He was making a summer tour of the garbage dumps of the Upper Peninsula and adjacent territory. There was, in truth, more to it than this and we will return to it again presently.

In the preceding days we had worked our way west from Hiawatha's river and Hiawatha's gulls. We had come through

Negaunee, where iron ore was first discovered in the Lake Superior region. We had skirted Mud Lake, where Lewis H. Morgan made his pioneer studies of the beaver. We had come to the Burnt Plains, where the repeated fires of the Indians long ago had produced widespread meadows for attracting deer. Beyond L'Anse, favorite camping ground of the early *voyageurs*, we had turned up the length of the Keweenaw Peninsula. For eighty miles this peninsula on a peninsula juts into Lake Superior, rising north and curving east above the western end of upper Michigan. It is unique among the peninsulas of the world. For down the middle of its entire length, like a backbone of metal, runs the world's most famous copper lode.

We followed the line of this lode northward into the so-called Treasure Chest of Michigan. Our highway was sometimes tinged with copper-red and, at intervals along it, immense heaps of excavated material rose beside gray buildings at the tops of mines. More than 8,500,000 pounds of copper have been produced here. One organization, the Calumet and Hecla Consolidated Copper Company, of which Louis Agassiz' son, Alexander, was president, has distributed approximately $200,000,000 in dividends.

Slabs of pure copper, carried by the glaciers as far south as Missouri, have been traced back to the Keweenaw Peninsula. Long before Columbus' ships were launched, Indians of this region were mining copper. Their ancient workings have left gashes in the earth that extend for as much as 500 feet. About twenty miles from the end of the peninsula, we passed a small stream. Along its course, centuries before, a waterfall had cut through a large vein of copper, providing the aborigines with easy access to the metal. At the far end of the peninsula, where Copper Harbor once formed the busy lake port of the boom period in the region, the lode dips down under Lake Superior to reappear again on Isle Royale.

Today Copper Harbor is a quiet, remote village visited

mainly for the beauty of its surrounding bluffs and shoreline. We reached it late in the afternoon, riding the high road over Brockway Mountain, wild, magnificent scenery spreading away around us. In a long plunge, the wall of an immense ravine dropped steeply down close beside our road. Far below, above the pigmy trees of the valley floor, we could see ravens diving, twisting, tumbling in the wind. Every valley amid this jumble of hills was a playground of the ravens. A fleeting rain, which stopped almost as soon as it began, sprinkled the road as we descended to the shore. It was the kind of precipitation Thomas Hardy once described as a "mild moisture which a duck would call nothing, a dog a pleasure, a cat a good deal."

In most parts of the country, communities carefully hide their garbage dumps. They ignore them as much as possible. They try to attrack attention away from them. But not on the Keweenaw Peninsula. And not in many other places in the forested north. We encountered numerous signs along the way, some even ornate, pointing out the route to the local garbage dump, as signs in other regions point "To the Park" or "To the Picnic Area." Instead of being ignored and concealed and held in disrepute, here they are advertised. They are tourist attractions. It is there that you see wild bears coming from the forests at dusk. In this land, garbage dumps are usually referred to as "bear pits."

That evening, in Copper Harbor, everyone seemed hurrying through supper. For here the biggest excitement of the day comes just when the day is over. It is when the twilight deepens that the bears appear and the tourists lose no time in getting in position to see the show. As soon as the meal was over, we heard motors starting and cars heading away toward the outlying refuse heaps. We joined the procession. Bumping along a rutted road, we came, in a few minutes time, to an opening in the woods. Already more than a dozen automobiles were ranged in a rough semicircle, all backed against the edge of the forest, all facing toward the disposal area—the

BEARS OF COPPER HARBOR 153

stage where the action would take place. Carefully I reversed, slid into a narrow opening between a gray car and a blue car, and cut off my motor.

A man from Minnesota sat at the wheel of the gray machine. In low-voiced conversation, we talked for a while as we waited for the twilight to deepen and the bears to come. It was under such circumstances that we made the acquaintance of the happy man. His great passion, he said, was wild bears. He wanted them free and uncaged. He understood their danger and kept his distance. He respected them as wild animals. But he liked bears, he admired bears, he never got enough of watching them. With undiminished enthusiasm and delight, he was in the midst of a grand tour of the garbage dumps of several northern states where bears were known to feed.

Apparently he was a charter member of that cult of avid devotees which we had hardly known existed—the bear watchers. Several were staying at Copper Harbor when we were there. We heard them in the lunch room. They compared notes. They reported on the largest and smallest bears they had seen. They debated over which refuse heaps, in which states, attracted the most bears. They recounted all the things they had seen the animals doing.

As the dusk increased, other cars arrived. The curving line kept extending at both ends. At last there were more than twenty cars—Cadillacs, Buicks, Fords, Oldsmobiles, Chevrolets, Mercurys, Chryslers, Plymouths, Dodges, a light truck and two foreign sports cars. The license plates represented a dozen states. More than fifty persons sat silent around us or talked in low tones. Only one, a woman several cars down the line, talked loudly. Her monologue, harsh and high-pitched, went on and on, as unpleasant as the sound of a dentist's drill. Twice local residents arrived with loads of garbage. Before they drove off with empty cans they stopped to look around at the lined-up cars. The animals are a commonplace; it is the assembled motorists that provide their show. They can't understand why people want to come and sit at a dump and watch bears.

Overhead a raven sailed by, croaked once, and disappeared. A little later a cawing flock of thirty home-going crows circled the clearing but made no attempt to land. Two impatient tourists, obviously not dedicated bear-watchers at heart, started their motors and went bumping away down the road. In the dusk, a child's treble voice called:

"Come on, bear!"

At last the bears came. The curtain had gone up; the show was beginning. This was what everyone had waited for, to see the wild beasts come shambling out of the forest shadows into the twilight of the open clearing. Few other creatures are as uncertain in temperament, as given to sudden change in mood, as bears. A wild animal trainer with a circus once told me that he could almost always tell what a lion or a tiger would do next, but he was never sure with a bear. On this night, only three of the unpredictable animals appeared. Sometimes, on other nights, twice as many came to feed.

But the show put on by the three satisfied the assembled spectators. There were grumblings, threats, sudden rushes and mounting quarrels over food. Two of the bears were about the same size. The third was a little smaller. It was the most timid of the three. Once when the woman with the dentist-drill voice stepped momentarily from her car with a folding Kodak and, in the deep dusk, snapped a picture that eventually she would see as a blank piece of film, this smallest bear fled to the woods. It was five minutes before it came back.

At some bear pits the animals come singly. The dominant bear feeds first. When he is done, the second most dominant bear comes to feed, and so on down the list. Here, all three bears came at the same time. But there seemed to be a pretty definite ursine pecking order among the three. The smallest and most timid was, of course, at the bottom. The two larger ones were fairly evenly matched, but one had an edge of superiority. It was the top bear. In any real contest it always was the winner. Once some choice morsel brought a clash between these two animals. The bickering grew louder. Threats and counterthreats rose in volume. For some time this quarrel continued, becoming more and more raucous. Suddenly the dominant bear made a rush at the other. It put up a brief resistance, then turned tail. The two, pursuer and pursued, disappeared down the road. A full fifteen minutes went by before they reappeared, first one and then the other, in the opening.

When the smallest bear discovered anything to eat, it never consumed it on the spot. It always carried it away from the other two to some private place and there enjoyed its prize in peace. All three of the animals used their keen sense of smell in hunting for food. Shambling along the front of the refuse piles, poking their noses this way and that, sniffing and snuffling, giving a mighty rake now and then with a forepaw, they moved from one end of the opening to the other.

Perhaps part of the greater respectability of these forest garbage dumps lies in the fact that they are better-smelling than those in most places. Over the clearing hung little if any odor of garbage, no smell at all of decayed food. Apparently, on their nightly visits, the bears were licking out cans, cleaning up scraps, discovering and rooting out all the food available. Guided by their keen noses, they were doing a thorough job of scavenging.

That summer, at one bear pit in northern Wisconsin, we were told, the caretaker occasionally buried pots of honey far down in the heaps of refuse. They never remained there more than one night before they were discovered by the noses of the bears. The proverbial attraction of honey for bears is so great that they will endure the stings of a swarm of defending bees to get it. In the fall of 1958, the New England Telephone Company had nearly fifty poles damaged by a plague of black bears in New Hampshire. The only explanation that the company could advance was that the bears heard the humming of the wires and, mistaking it for the sound of bees, had been trying to find the honey.

Everything from ladybird beetles and ants and fish, through berries and bark, grass and roots, mice and frogs, to carrion and honey is included in the diet of the omnivorous bears. The man in the gray car, who was happily watching all that was taking place, told us afterwards that at Three Rivers, Wisconsin, one of the biggest bears he had ever seen—he thought it must have weighed more than 400 pounds—stopped rooting in the refuse piles and began looking fixedly in his direction. Then it turned and shambled directly toward him. It kept coming until it reached the front bumper of his car. Here it stopped, bent down and began licking the chromium of the grill. It was getting mayflies!

Of all the food bears consume, the dainty mayfly seems the strangest. Yet, oddly enough, these great beasts have a special fondness for these airy, diaphanous insects. Ernest Thompson Seton, in his *Lives of the Game Animals*, tells of black bears in Canada that visited the edges of far-northern lakes to feed on the mayflies. He watched them eating ravenously, scooping up masses of the dead ephemerae which had collected in piles and windrows along the shore.

In the late 1930's, when watching bears at disposal dumps began to attract tourists on the Keweenaw Peninsula, officials were presented with a double problem: first, how to protect the people from the bears and, second, how to protect the bears from the people. By a program of education, emphasizing that the animals were wild, untamed and potentially dangerous, they induced visitors to remain in their cars. Then, before the deer hunting season began, when all the bears that had become accustomed to come to the dumps in safety would have been slaughtered by the gunners, special legislation was enacted protecting the animals. As a result, on this evening, the bears were there and all the spectators kept their distance and watched events from the grandstand seats of their automobiles.

They remained there until the foraging animals became dim shapes moving in dusk that was almost darkness. For a time, someone in one of the cars ran the small beam of a powerful spotlight over the feeding animals. They were not disturbed in the least as they rooted about, seeking the last of the food. When that was found and consumed, they turned away one by one, padding off into a forest that now was filled completely with the blackness of the night.

When the next evening fell, it found us beside a little bay of Lake Superior, west of the Keweenaw Peninsula, not far from the Porcupine Mountains, and a hundred miles from the bears of Copper Harbor. We had moved west along the shore and through the green of the forest. Summer had reached the time that Vergil described so long ago: "The woods are now in full leaf, and the year is in its highest beauty." On this day, Nellie had seen a bird she had been seeking through all the Upper Peninsula, the mourning warbler. We had looked down on the Lake of the Clouds amid the Porcupine Mountains, the highland of Michigan. And, sometime after sunset, we had come to this quiet bay on "the shores of Gitche Gumee."

The sand that sloped gently downward was strewn with flat, water-smoothed stones of many hues. We wandered aimlessly among the pebbles, relaxed and enjoying the tranquil ending of the day. In a long return to that almost universal pleasure of boyhood, I began skipping flat stones across the tinted water. Gustave Flaubert, the French novelist, once observed that even the simplest thing contains a little that is unknown. I wondered idly at the source of my pleasure. Did it arise entirely from the game, the competition with myself, the exercise of skill? Or was there in it the thrill of producing an effect that, for a time, defeated logic and defied gravity—supporting stones on insubstantial water? I never did decide. And as I warmed to the age-old sport, I soon quit trying. The number of skips rose. What did the reason for my enjoyment matter? At the time, just to enjoy it fully seemed enough.

SIXTEEN

FARTHEST NORTH

RAVEN roads, with the great black birds flapping up in the dawn light, carried us west from the Porcupine Mountains next morning. Before nine o'clock all of Upper Michigan, that land of wonderful wildness, lay behind us. Along the topmost edge of Wisconsin, we followed the Superior shore through Ashland, on our right rust-red water tumbling in the wind. Like a beetle on a saucer rim we crept around the lake's far western end. We passed the great sand bars that form the protected harbor of Superior and Duluth. Then we were running north—past Floodwater Bay and Gooseberry River and a place the Chippewas once named Spear-by-Moonlight.

The water beside us lost its sediment-tinting of red; it grew richly, deeply blue. We looked back as we swung west away from the shore at Baptism River. Between the trees we glimpsed for a final time the waves and the sparkle of the last of the Great Lakes. Since before Niagara Falls we had been following the shoreline of these five inland seas. They had been companions of our travels almost from the beginning. Now we were heading west, the five Great Lakes behind us, a land of 10,000 little lakes before us.

Scattered across the 84,000 square miles of Minnesota, there are well over that number of small bodies of water. They represent almost one-tenth the total inland water area of the United States. Here is a Silent Lake and a Sleepy Eye Lake, a

Little Dead Horse Lake and a Full of Fish Lake, a Bowstring Lake and a Disappointment Lake and a Mantrap Lake. More than 100 names are repeated five times or more and the state contains ninety-nine Long Lakes, ninety-one Mud Lakes and seventy-six Rice Lakes. In this canoe country of the north it is estimated there are fully a thousand lakes still unchristened. We amused ourselves that afternoon by conjuring up names for the unnamed lakes of Minnesota.

Many times that day our road reached away before us like a corridor walled in by the forest. As we advanced, aspen trees, the popple or beaverwood of the north country, increased around us. The aspen, belonging to the poplar family, is one of the most widespread of all American trees. It ranges from the arctic down the western mountains almost to the borders of Mexico. Of all the tens of thousands of these trees we saw that day, one stands out in memory. A lazy breeze had turned a hundred or so of its leaves silver-side-out. They shone in the sun, scattered like white butterflies across the green of the foliage.

The glinting dance of the poplar leaves, now dark, now light, adds life to any landscape on a windy day. Nearly ninety years ago, the young English clergyman, Francis Kilvert, whose recently published *Diary* records his intense appreciation of nature, found special delight in the movement of these leaves. "For some time," he noted in his entry for Wednesday, October 7, 1874, "I have been trying to find the right word for the shimmering, glancing, twinkling movement of the poplar leaves in the sun and wind. This afternoon I saw the word written on the poplar leaves. It was 'dazzle.' The dazzle of the poplars."

Across the top of Minnesota, above the source of the Mississippi, through swamp and farm and forest land, we advanced during the following days. Twice along the way we saw high piles of gravel beside the road. In the sheer face of the cut-away side of each pile, bank swallows had made their

FARTHEST NORTH

nests. We passed International Falls and beside the Rainy River we worked west along the Canadian line toward Lake of the Woods and the Northwest Angle. This thirty-mile wedge of land and water owes its existence to the mistaken belief that it contained the source of the Mississippi. In 1783, at the Treaty of Ghent, which recognized its independence, the United States insisted on this jog in the Canadian boundary. At no other place along the border does American territory extend north so far. The tip of the Angle was for us the farthest-north point of our summer trip.

More than a hundred years ago, an English traveler, George W. Featherstonhaugh, published in London an account of his travels in the wilds of Minnesota. The title of his book was: *A Canoe Voyage up the Minnay Sotor*. The century of mechanical advance that intervenes has brought many changes to "Minnay Sotor"—both river and state. But in the Northwest Angle the wilderness still presses close. This roadless land, without complete topographical maps to this day, contains even now stretches that represent the forest primeval.

It was late in the afternoon when we came to America's only port on the Lake of the Woods, Warroad on Muskeg Bay. The name commemorates the old war path followed by the Sioux and Chippewas around the southern end of the lake. Warroad, occupying a till-plain where the waves of ancient Lake Agassiz leveled the drift deposits of the Ice Age, seems only a day or two removed from pioneer times. Still legible on the front of a weathered building is a large sign: "Homesteads Located by an Actual Settler."

We slept that night beside Muskeg Bay, and early the next morning, aboard the veteran *Resolute*, we were on our way to Angle Inlet at the far tip of the wedge. Ever since it was built on the lake shore shortly after the First World War, the sixty-two-foot *Resolute*, with Fay H. Young as owner and captain, had been threading its way among the islands each summer. And each winter it had been frozen in ice that sometimes reached a depth of six feet, carefully anchored heading north and south so the melting in spring would occur equally on either side and in this way eliminate the danger of capsizing.

A laboring diesel, one of its cylinders inactive from a burned-out piston, pushed us north past Buffalo Point and Springsteel Island, out into the width of Big Traverse Bay. With more than 200 square miles of water surface, partly in the United States and partly in Canada, the Lake of the Woods is sown with rocks and reefs and innumerable islets. It is estimated that there are 14,000 islands in the lake. These, added to the convolutions of the lake rim, are said to give the Lake of the Woods more miles of shoreline than any other body of fresh water in the world.

Storms were predicted for the day. The sky was dark and the wind was rising as we entered the Big Traverse. For thirty miles here the wind sweeps unhindered, piling up running ridges of water in rough and sudden storms. This day the long gusts came out of the east. The moving moraines of the waves swept toward us across the lake, striking the *Resolute* broadside or at a quartering angle. Clinging to the rail, we would see the opposite rail drop down and down until it almost touched the water. Then, as the boat slid into the trough and the next great wave reached us, it rose, soaring upward as we fell away. Thus, rising and falling on our aquatic teeter-totter, we crept slowly northward.

Fay Young, our seventy-year-old captain, was a fine man to have around in a storm. We could see him peering ahead through the small pilothouse windows. The throb of the diesel rose and fell, easing off on the water ridges and digging in during the momentary lull of the troughs that followed. At the end of each steep slide-down, a white burst of spray would explode upward around the bow. Once a geyser of foam shot over the top of the pilothouse. I could see Fay Young cranking the hand-operated windshield wiper to clear his windows. Across this seemingly endless procession of waves, each time the boat heeled over the bell above the pilothouse would clang once dismally, as with a tolling sound.

"This is the kind of a day you'd like to forget!" In these words the captain summed up the trip succinctly.

Only once in his forty years on the lake, he said, had he had a rougher crossing of the Big Traverse. That occurred during a near-hurricane one June several years before. On our trip, the particular yawing and slewing motion of the boat, combined with its rearing and pitching, left almost all of the dozen passengers violently seasick. Even dogs sometimes become seasick on the Lake of the Woods. Fortunately for us, Nellie had remembered a bottle of Dramamine pills. Fortified by one apiece at breakfast time, we rode out the storm hour after hour.

For a long time I clung jammed between the rail and a red-painted davit near the stern, with the spray and the wind around me, elated by the constant change, the shifts in equilibrium, the sweep up and down, this way and that, the boat continually righting itself, conquering wind and wave with buoyancy and equilibrium. Here was no steady, bumping advance of the laden wagon, no smooth, metallic run over rails. Here was fluid motion, the stability of instability, the sensation of canoe or bicycle multiplied manyfold. Here was the excitement of the lake in a storm.

All around were the steep hills of water with the snow of foam on their ridgetops. They seemed devoid of life. At long intervals a Franklin's gull or a black tern beat past us against the wind. No longer are there in these waters the incredible numbers of sturgeon that once swam here. In the single year of 1895, for example, 1,300,000 pounds of these fish were taken from the Lake of the Woods. The roe alone totaled 97,500 pounds. One giant sturgeon caught in the Big Traverse weighed 285 pounds. But the prodigal waste of an earlier time has left this natural resource depleted. The sturgeon is now ranked in the Lake of the Woods as "very scarce."

For nearly four hours the *Resolute* lurched and plunged in heavy water crossing Big Traverse Bay. Then the lake grew calmer, protected by the maze of the upper islands. At each port of call, the captain was greeted enthusiastically. To every salutation of:

"How're you, Fay?" he would mumble:

"I managed t' get this far!"

Then, with cargo discharged, he would swing away with a farewell salute—a feeble wheeze of the boat's hand-cranked whistle—and we would head north again. We were to see it all once more, this time in brilliant sunshine, when the boat headed south again next day. Now on a water frontier we advanced toward wilder land ahead. The lowering day was well along in its fourth quarter before we reached the end of our journey, Jake Colson's wharf and the few scattered cabins that form the community of Angle Inlet. Here all the traffic goes by water. The school bus is a boat. Inland, the muskeg bogs form impassable barriers except when they are solidly frozen in winter.

Three miles farther on, on the banks of the dark Bear River, at the very tip of the Northwest Angle, Lockwood Jaynes took us in for the night. A metal worker from Tulsa, Oklahoma, he and his wife have spent their summers here for years. Nearby he was building a new reinforced and insulated cabin. There they planned to live the year around, the only dwellers on this wild northland stream.

Dark brown with the leachings of the Manitoba bogs where it originated, the river moved slowly, its farther bank a hundred feet away. The main current wound back and forth along its wider bed, each meander in its serpentine advance holding within its curve shallows green with wild rice. For two hours or more in the late afternoon and the long northern twilight, Nellie and I explored upstream, almost to the Manitoba line, pushing a heavy rowboat along the windings of this river within a river.

FARTHEST NORTH

At first I tried to take short cuts across the rice shallows. But it was slow, laborious work. Among the multitude of the rice plants—their seed-stalks upthrust, their green leaves extended on the water—my oars became entangled at every stroke. Each time we neared the shore, as we wound with the channel, we seemed rowing in the midst of windblown trees. Disturbed by the waves and ripples my oars produced, the reflections of all the streamside aspens shortened and lengthened and swayed on the dark mirror of the water around us.

At the quiet end of this overcast day, now without wind, the river seemed tame enough. It reminded us of stretches of the sluggish Concord in Massachusetts. We kept expecting to come upon clearings and farms, to see cows and barns. But the only sign of human habitation we saw was the moldering ruins of a log cabin, built long ago, now with roof collapsed and overgrown with moss decaying like a fallen tree and being reabsorbed into the forest. Only a few nights before, Jaynes had heard timber wolves howling. And hardly a mile inland from the stream the traveler is turned back by a virtually impenetrable barrier of tangled forest and treacherous muskeg.

Beaver-felled alder and aspen sticks, stripped of their bark, floated past or lay stranded in the rice shallows. At almost every turn we came upon the remains of beaver houses. Once, as we rounded a bend, we surprised a doe more than half submerged in the stream, escaping the flies. All over the north country at this time of year the larger animals suffer from the multitudinous insects. On the boat going back to Warroad next day, an old-timer told us of coming upon a moose in the deep woods in midsummer. Even before he saw it, he said, he detected its presence by the buzzing of the flies, the sound so loud it reminded him of "an airplane in the sky."

For three days before our arrival the weather had been abnormally warm. Thermometers had risen almost to 90 degrees. Natives spoke of this as their "three-day summer." And all along the river, in the short blooming season of the north, water plants had opened their flowers. We drifted by the white globes of the bur reeds and the white, three-petaled blooms of the arrowheads. We floated among the crowfoot, that aquatic buttercup with flowers of white instead of yellow. Among all the river blooms white predominated. Chief exceptions were the yellow pea-like flowers of the bladderworts and the buds and open flowers of the yellow water lilies. Where a beaver trail led away into the woods, one small bay seemed plated all around its edge with the gold of the lilies.

As we were drifting back downstream, we stopped to examine a green rush thrust above the water and coated for half a foot with an almost solid mass of aphides nearly the same identical shade of green. A little farther on, at the outer edge of a quarter of an acre of wild rice, we discovered a large gray longicorn beetle struggling in the water. I fished it aboard on the end of my paddle. For a time, like Fabre's pine processionary caterpillars, it went round and round the rim of my hat. We headed for the bank to toss it ashore. And in so doing we encountered a special little adventure of our own. For the beetle led us to a family of kingbirds during one of the most eventful hours of their lives.

Ten or eleven feet from the ground and a dozen feet from the edge of the water, a pair of these birds had made their nest where the dead stub of a branch jutted out from the trunk of an aspen tree. Five baby kingbirds, almost ready to fly, were lined up on the stub. They clamored continuously for food while the parent birds flew back and forth bringing insects. We saw a large grasshopper stuffed hastily down the gaping throat of one fledgling. Each time a parent appeared the calling rose to a crescendo. All the young birds crowded together. They opened wide their orange-yellow mouths. Their outspread wings quivered in that age-old action of supplication and hunger instinctively used by birds around the globe.

Each time the parent left, the fluttering of the wings subsided. But there came a time when, as we watched, the fourth fledgling from the end of the stub continued the movement. At first the outstretched wings pulsated as before. Then they fluttered faster and faster. The arc of their movement increased perceptibly. Suddenly it was in the air. It was launched into flight, rising to another branch two feet higher up in another tree. From the quiver of the hunger movement to the beating of wings in flight had been a smooth transition.

A little later a second fledgling similarly took wing. Then a third broke home ties, alighting in a neighboring aspen. Finally, before we drifted away downstream, a fourth had launched itself into the air in exactly the same manner. Each rose to a higher perch. Each translated the flutter of hunger into the more frenzied wing beating of the first flight.

This was something I had never realized before. It is something I have never seen mentioned in ornithological literature. The wing-fluttering apparently becomes more than a symbol of hunger for the parents. It represents something in addition to strengthening muscles of flight. It forms a vital link in the chain of instinctive behavior that leads the bird into its aerial life. In his classic experiments with the Sphex wasps of his Provençal countryside, J. Henri Fabre demonstrated how the activity of these insects forms chains of instinctive behavior, with each act in the sequence linked with and leading into the next. So among these baby birds one act merges with another in this mechanism of nature for breaking a fledgling's ties with the nest.

We were half a mile from the cabin in the still evening air when a sound like wind rushing through the trees grew louder. Immense raindrops, each producing a gleaming bubble on the dark surface of the stream, fell all around us. In this pelting shower we reached the dock and the shelter of the cabin. The rain drummed overhead as we ate with the Jayneses a late supper of homemade bread, fresh vegetables from a wilderness garden and fruits of the forest, wild strawberries and raspberries and blueberries.

Outside, on the steps of the cabin, the deluge was washing away the little piles of dandelion fluff and the scattered pulp of wild strawberries, both discarded by the chipmunks as they nipped away the seeds. While the rain without wind continued, then slackened, then ceased altogether—to be succeeded by the dripping of innumerable leaves—we talked of the Old Dawson Trail, the overland route to Winnipeg over which Red River ox carts had labored in pioneer times. Traces still remain to the west of Angle Inlet. Aside from these, the forests are largely trackless, a waste where a man without a compass would soon become lost.

Even with a compass, Jaynes recalled with amusement, he once lost his way while hunting in the fall. Carrying his gun under his arm, he pulled his compass from his pocket from time to time and followed its direction. Suddenly he realized, as he said, he was nowhere near where he thought he was. He awoke to the fact that the steel of the gun barrel was deflecting the compass needle. Hurriedly he shifted the rifle to his other arm and began retracing his steps.

How completely confused even an old inhabitant can become in this region was illustrated, one recent winter, by a man who had lived near Angle Inlet for nearly forty years. While hunting on a heavily overcast day he discovered he was lost. For hours he wandered about before he came to a house in a clearing. He was just about to go in and ask where he was when he realized the house was his own.

The emotion of being lost, of being helpless, of being in a topsy-turvy world where everything is unreal and unfamiliar, often produces terrifying psychological consequences. At times men have lacked the wit to trace back their own trails in the snow. In the north woods of Maine I heard of one hunter who passed a group of other hunters on the run. He shouted: "I'm lost!" Then he disappeared in the forest, still running at top speed. Not infrequently guns and clothing are thrown away. In the rain forest of the Northwest, I was told of an elk hunter who was trailed for two days by a rescue party. They found garment after garment discarded along the wandering path he had followed and when they came upon him, at last, he was standing by a tree clad only in shorts.

It is Ken Slater's idea, and it seems to me a good one, that a few simple instructions for lost persons should be printed on the back of every hunting license. Government maps of national forests now carry a list of suggestions on what to do if lost, such as: Stop. Try to figure out where you are. Do not yell. Do not wander about. Do not run. Travel only downhill. On several occasions, in recent years, the far-carrying racket of a chain saw has been used to guide lost persons out of the forest.

Before we went to bed late that night, at the end of this full day, at the far northern point of our summer journey, we walked for a while beside the dark river. Along our path, over the water, at the edge of the dripping forest, fireflies floated. Unlike those living sparks of the Kankakee, they neither danced nor shot upward into the air. They drifted slowly, large, glowing, greenish lamps in the velvet darkness. When we came in at last we traded the rich perfumes of the rain-soaked forest for the nostalgic smell of the kerosene lamp in our cabin room. Long after its yellow glow was gone we lay listening to the sounds of the darkness of this northland night, the hoot of a great horned owl and the slow grating call of frogs in the river shallows.

SEVENTEEN

THE WATER PRAIRIES

IT was now late in July. Midsummer had settled over the Great Plains. Almost 300 miles west of the Lake of the Woods, just below the Canadian border in North Dakota, a little more than half way from Minnesota to Montana, we had reached the great loop of the Souris River. Meandering down across the border from Saskatchewan, the Souris curves to the east and then turns north, flowing away into Manitoba. Its course marks roughly the lower boundary of a prehistoric glacial lake that was left in the wake of the last ice age.

When the first French trappers penetrated to this marshy region, they found it teeming with waterfowl and all the higher land swarming with mice. They named the stream the Souris, or Mouse River. In 1800, four years before the start of the Lewis and Clark Expedition, the English fur trader, Alexander Henry, visited the Souris. He noted in his journal: "The mice destroy everything; they eat my skins and peltries—indeed anything that is not iron or steel goes down with them."

In the intervening years the mice have become fewer. The trappers and fur traders were followed by homesteaders and land speculators. In an ill-advised attempt to turn all the area into farmland, the great marshes along the river, the immemorial home of the waterfowl, were drained. In 1935, when the failure of this enterprise was apparent to all, the U.S. Fish and Wildlife Service began the task of restoring the river bottom to marshland. By impounding water upstream

THE WATER PRAIRIES

and letting it descend at intervals to replace losses by evaporation, wide areas are kept flooded behind a series of dikes. The result is one of the most important migratory-waterfowl refuges in America. At times around us, the immense marshy ponds seemed to stretch to the horizon. They were the water prairies, level and outspread beneath the high blue of the Dakota sky.

They were all the more impressive in a land of lost lakes. Coming west through upper North Dakota, across the great swells of the open land, we had passed many dry beds of former ponds. Within the space of twenty years, it is said, the state saw one-third of its small bodies of water disappear. Originally the prairie sod held the rain and reduced the dissipation of moisture into the dry air. As plowed fields extended over an ever-greater area, they increased the rate of evaporation. Moreover, a swiftly multiplying number of artesian wells have increasingly drained away water resources that underlie the state.

The Dakota Artesian System, extending west into Wyoming, east into Minnesota and south into Nebraska and Iowa, is one of the most famous in the world. Slanting downward from outcroppings on the flanks of the Black Hills, the Big Horn Mountains and the eastern Rockies, a vast sheet of gray, porous Dakota limestone is overlaid with a stratum of almost impervious shale. Water percolates downward along the sheet of limestone as a stream descends its bed. It builds up underground pressures that provide flowing wells whenever the shale is penetrated and the limestone reached. The first such well in the world was sunk in a convent yard in the province of Artois—from which the name artesian is derived—in the south of France. It has flowed without ceasing since 1126. It was seven and a half centuries later, in 1882, when the Dakota Artesian System was discovered.

At first, pressures were as great as 250 pounds to the square inch. They shot columns of water to heights rivaling those of the geysers of Yellowstone. In some instances they were harnessed to run flour mills. Steadily the number of wells tapping the underground water supply increased. There were 400 in the two Dakotas in 1896, 5,000 in North Dakota alone in 1921 and more than 15,000 in the two states in 1930. At first it was maintained that the underground supply was inexhaustible. But by the turn of the century a decline in pressure was already being noted, and by 1916 laws were passed restricting the use of artesian water. In 1923, when another well was sunk to the same depth that, in 1886, had produced a flow of 600 gallons a minute at Ellendale, North Dakota, the water rose only flush with the surface and did not flow at all.

At the edge of Devils Lake, the largest natural body of water within the boundaries of North Dakota, we had stood beside a high pole that recorded with dramatic simplicity the lowering of the water level in the region. A full twenty feet above the ground, a marker indicated where the surface of the lake had stood in 1870. By 1880, it had sunk four feet; by 1900, it had dropped thirteen feet; and by 1910, it had fallen sixteen feet. When, in the 1880's, the town of Devils Lake was established on a low glacial moraine close to the shore, a steamboat landing was constructed on the outskirts of the community. By 1920, the edge of the lake had receded two miles from town. Today it is five miles away. With this background of receding water in our minds we came, with added interest, to the outstretched ponds of the Souris refuge.

Moving along the dikes, on foot or in our car, we were always surrounded by dragonflies. Once I counted fifty, darting, whirling, hovering, advancing as we advanced, surrounding us with a cloud of shining bodies and glittering wings. We heard the faint clash of their parchment-dry wings on vegetation, a sound older on this earth than the song of any bird. When we slowed to a crawl, this insect convoy thinned out. As soon as we stopped and stood still for any length of

SHOWY LADY'S SLIPPER, delicately perfumed and daintily colored, blooms in a boggy stretch on the Door Peninsula.

RED SQUIRREL at the Seney National Wildlife Refuge on Michigan's Upper Peninsula. Below, a least chipmunk.

time, the cloud dissipated entirely. Swishing through the grass in our advance, we were sending innumerable small insects fluttering upward into the air. The dragonflies had assembled for the feast.

Occasionally swallows, too, darted close about us. Among the 253 species of birds that have been recorded in the vicinity of the river, five different kinds of swallows nest within the boundaries of the refuge. Only here, in the marshes of the Souris, is it possible to find the nests of all five American grebes—the red-necked, or Holboell's; the pied-billed; the eared; the horned and the western. When we were there, for some reason, all grebes were scarce. We saw but few. But a multitude of other birds made up for it. One hundred and nineteen species—water birds, shore birds, songbirds, game birds—are regular summer residents. Here our first sharp-tailed grouse went whirring away, tilting or rocking from side to side in characteristic flight. And here we encountered our first LeConte's sparrow, flitting low over the grass tangles on the inland side of the dikes or pausing at intervals to repeat its thin and insectlike little song. On our first day at the Lower Souris Migratory Waterfowl Refuge we added twelve new birds to our summer record and three to our life lists.

We came upon the third of these life birds early in the afternoon when Donald V. Gray, the energetic and helpful Refuge Manager, showed us the immense variety of the sanctuary—the miles of riverbanks and dikes, the wooded bottomland, the sandhills and the rolling short-grass prairie. For more than twenty-five miles, the refuge follows the river northward until it reaches the Manitoba line. Once we came upon a badger—big, gray, low-slung—moving on short and powerful legs like a thick rug sliding over the ground. In several places, our companion steered around badger holes in the road. By preference, the animals frequently dig their burrows in the hard-packed earth of the wheel tracks. At times we surprised deer in their feeding. For some reason that, so far as I know, is not entirely clear, the white-tailed deer grow larger as you go west. In Minnesota they are slightly larger than in Michigan, and in North Dakota they are larger still. Here they reach their maximum size and weight.

Somewhere in the northern part of the refuge, Gray pointed out a stretch of virgin sod. Our road shimmered away into the distance, two tracks across the yellowing grassland, paralleling the flow of the river. We stopped at a gate and noticed, fifty or sixty feet beyond, an active flock of little birds feeding in the wheel tracks, beautiful birds we had never before seen in our lives.

The females were brownish, striped and sparrowlike. But the males, still resplendent in their breeding plumage, were black-capped and all black beneath, with a wide band of chestnut-red extending around the nape of the neck. They were those birds of the open plains, the chestnut-collared longspurs. We watched them hurry about over the packed earth, snapping up some small seed or insect, then hastening on. Most birds take a rest or siesta during the midday heat of summer, but not these longspurs. One of their characteristics is that they continue foraging all through the hottest hours.

Half a mile farther along the wheel tracks, we stopped beside an odd, polished boulder of pinkish rock resting at the center of a bowl-like depression thirty feet across. The rock was about three feet high and five and a half feet long. It reflected the rays of the sun, shining with a high polish. We ran our hands over its smooth, burnished surface. Across all the prairie land of this region, herds of bison once had roamed. The boulder is one of four rubbing rocks—"buffalo rocks"—discovered on the refuge. Each lies in a similar depression. There, through the centuries, the hoofs of the animals cut through the grass around the stone, left the loose, trampled soil exposed to the wind, and thus produced these basins on the plains. How many hundred thousand times, we wondered, had the bison rubbed against the sides of this rock to produce, in the span of how many years, this high luster that caught the sun?

Cool nights and cool dawns, hot noons and hotter afternoons are the rule of the North Dakota summer. The coolness of one evening was beginning when Nellie and I came to a wide stretch of open land extending away toward the bright glow of the western sky. Scattered across it, upland plover fed, flew and landed, each time stretching high their slender, graceful wings before they folded them at rest. For some time we watched them. At first all the rather small heads, set at the tops of slender necks, were lifted to study us. A few moved their heads forward and backward horizontally several times, to see from different angles, as they scrutinized us intently. But soon these large inland sandpipers were at ease. They wandered about, peered down, snapped grasshoppers from the ground and vegetation. As we watched them toward the luminous west, they moved amid the shimmer and shine of summer gossamer. For all across the open land silken threads, where ballooning spiderlings had landed, glinted in the last bright light of the day.

Now and then one of the feeding shore birds called. We talked about how different are the emotions produced by the varied voices of the birds—the mellow, relaxing, slumberous notes of the mourning dove; the wild, stirring call of Canada geese in the sky. The voice of the upland plover is infinitely moving. It is rich and mellow, with overtones of sadness in it. In flight, its voice swells in a clear, far-carrying, mournful whistle that fades away in the sky. W. H. Hudson, in some of the last pages he wrote, when he was more than eighty years old, recalled the impression it made upon him when he heard it as a boy on the pampas of Argentina.

"Lying awake in bed," he writes in *A Hind in Richmond Park*, "I would listen by the hour to that sound coming to me from the sky, mellowed and made beautiful by distance and the profound silence of the moonlit world, until it acquired a fascination for me above all sounds on earth, so that it lived ever after in me; and the image of it is as vivid in my mind at this moment as that of any bird call or cry, or any other striking sound heard yesterday or but an hour ago. It was the sense of mystery it conveyed which so attracted and impressed me—the mystery of that delicate, frail, beautiful being, traveling in the sky, alone, day and night, crying aloud at intervals as if moved by some powerful emotion, beating the air with its wings, its beak pointing like the needle of the compass to the north, flying, speeding on its 7,000-mile flight to its nesting home in another hemisphere."

Here on this northern plain, in that other hemisphere, we were seeing the birds that had reached their nesting home. And already, now, the long flight back was beginning. Some appeared to be young birds. They were a little fuzzy about the head as though faint remnants of down remained to be molted. Along the edges of the water prairies, we saw other shore birds gathering into knots or groups. In North Dakota, birds begin forming migratory flocks as early as July 10. Sometimes in the upper part of this state, migrants pass each other going in opposite directions. The earliest nesters are already moving south while the latest nesters are still working northward toward their breeding grounds.

It is, of course, during the main periods of migration in spring and fall that the great ponds of the Souris refuge are most alive with waterfowl. But even with only the summer residents floating on the open water or swimming away among the cattails and rushes, we were among widgeon and teal, pintail and shoveler, mallard and redhead, canvasback and scaup, gadwall and ruddy duck. From the bobolinks of the hillsides to the Wilson's phalaropes landing together a hundred at a time on the surface of the water; from the marbled godwits stalking along the pond edges to the white pelicans and sandhill cranes sailing overhead, the Souris refuge forms a summer-long paradise for the watcher of birds. It was with deep regret that

THE WATER PRAIRIES

we came to our last day there. I remember we stayed out that evening without supper, remaining until it was too dark to see more.

Standing on the dikes after sunset, with the expanse of the water prairies reflecting the fading glow of colors in the sky, we watched the waves and rivers of home-coming birds flowing overhead. Franklin's gulls came in, flock after flock, stream after stream, thousands upon thousands alighting at the far side of the pond and whitening the water. In one huge trailing V, fifty-four white pelicans sailed above us and, without the movement of a wing, soared on and on to reach their lodgings for the night, a remote bar at the farther end of the impounded water. Steadily, in the fading light, black terns streamed by, all heading south, out over the floating waterfowl, toward a stretch of rushes and cattails that extended in a low, dark line along the horizon.

A little way out from shore, one of the few eared grebes we saw swam slowly along, a baby grebe riding on her back. Suddenly the mother bird dived, spilling off her passenger. A moment later she reappeared with a small minnow in her bill. This she fed to the little grebe. It gobbled it down, then rushed about like a frisky colt, once leaping clear over its mother's back.

We turned from this amusing spectacle to look inland. There an arresting sight held us motionless. All across the grassy lowland towering columns of pale dancing light rose a hundred feet into the air. They expanded, merged, drew apart, joined together into one vast, shimmering, living northern lights. The Souris midges were rising into the quiet air of evening in numbers to stupefy the imagination. Their pale yellowish, almost translucent bodies seemed to glow as they caught and reflected the last light of the day. Behind their billion whirling bodies, all the landscape wavered as though seen through dancing heat waves.

I walked down the dike-side and out into the grassy expanse. Nellie watched through her binoculars. Signaling like surveyors, we changed my position until I stood in the very midst of the largest column. The air seemed half solid. I could feel immense numbers of almost weightless insects pelting against my face and hands. They covered my clothing. They piled up in the eyepieces of my binoculars. The noise of their wings, individually inaudible, collectively rose and fell in humming waves of sound. After nine o'clock, when stars were visible and short-eared owls had begun their hunting, the mating dance continued. Even when the forms of the insects were entirely lost in darkness, we could still hear the sound of their multitudinous wings. Once more, as among the Lake Erie mayflies, our minds were overwhelmed and staggered by the fecundity of summer life.

EIGHTEEN

BETWEEN HAY AND GRASS

THE first western kingbird called at 4 A.M.; the first robin at 4:30. Soon afterwards we were on our way. Running north we saw the flame hues of the prairie dawn reflecting in brilliant flashes of color from the windows of distant, silent farmhouses to the west. Just below the Canadian boundary, we turned away from the sunrise on State Highway 5. It carried us below Manitoba, below Saskatchewan, west to Montana. Amid great sweeps of rolling, dun-colored hills, with permanent snow fences beside the road and tiny horses feeding on a distant skyline, we traveled the final North Dakota miles. At 10:44 that morning, we reached the Montana border and entered the fifteenth state of our travels through summer.

The highway before us now rolled up the hills and down the hills, stretching straight toward the town of Plentywood. The name of this community in the far northeastern corner of Montana had an ironical ring. For on all sides the parched and treeless hills seemed skinned. We were on the high plains. And, when we reached it, Plentywood, belying its name, exhibited no more than a few scraggly poplars and box elders gray with dust. At the Montana line, we had crossed into Mountain Time, gaining an hour. Instead of noon, it was eleven o'clock and Plentywood restaurants were still serving breakfast. But the menu was a substantial one. It offered such dishes as one hitherto unencountered in our wanderings—fried-oyster sandwiches for breakfast.

Fortified by more conventional fare—scrambled eggs and bacon—in this breakfast-at-midday, we turned south, beginning a long zigzag down the eastern portion of Montana. We were riding across open land—open to the blaze of the sun, shimmering in the heat waves, with no tree, no shelter, no hiding place anywhere. In mounting heat we rode through the village of Antelope. Rangeland gave way to immense wheat fields separated by darker strips where land lay fallow to store up moisture. The long rectangles of grain stretched to the horizon. Whole mountains seemed clad in alternating strips of gold and green or dark green-gold. Never before had we been in the midst of so vast a sea of grain.

Everywhere the summer harvest was in full swing. Red combines, cutting and threshing the grain in a single operation, moved about the fields, each in a whirlwind of chaff. All along the roads of eastern Montana we met the parade of free-lance combine operators hurrying from job to job, their harvesters loaded on trucks with, oftentimes, a second truck pulling a house trailer behind. For the owners of the wheat fields, this was harvest time, journey's end for the year. Like ship captains, with storms behind them, they were coming safely into port.

Before we descended into the valley of the upper Missouri, near Culbertson, wheat farms had disappeared. The soil had become thinner, drier, breaking through the sparse vegetation here and there in reddish patches. Later in the afternoon, in ever-mounting heat, we paralleled an immense curve of the Yellowstone River on our way to Glendive. Wherever we looked the buttes, the grain fields, the sagebrush shimmered in the heat waves. Everything pulsed and vibrated. Nothing, even at a comparatively short distance, seemed solidly planted or entirely stationary. Thus we rode south through an unsteady landscape. Our noses seemed parched. Our eyes felt hot. Our heads throbbed dully. The desiccating wind appeared to suck all the moisture from our bodies. This was our first big heat wave.

Birds along the way were mostly silent, mostly out of sight. Infrequently we saw a meadow lark—pale-hued, the color of coffee that is half cream—perched on a fencepost, its mouth open, its wings drooping. If we had been birds, we decided, our wings, too, would be drooping. If we had been plants, our leaves would have been wilted. It seemed to us the hottest day we could remember. And so it was. For when we reached Glendive all the thermometers stood at 105 degrees F.

The raw, bare bluff just across the Yellowstone projected and magnified the heat. Glendive is a river-valley town with built-in reflectors. According to the American Guide Series volume on Montana, the temperature at Glendive, on July 20, 1893, reached the highest point ever recorded for the state, 117 degrees F. In the shade of a red-brick freight house, late that afternoon, I talked to an elderly man who remembered that day. He was a small boy then, tending sheep. Usually the flock went down to the river to drink about noon, he said. But on this morning they all started for the water at nine o'clock. Nothing would make them budge the rest of the day.

Like those sheep of more than sixty years before, the people of Glendive were all lying low on this day. We found a place to stay and ran cool water over our wrists—one of the quickest ways we have found for reducing the temperature of the bloodstream and finding relief from summer heat. The flaring sundown had faded away over the broken hills before people began emerging into the evening air. The dusk cooled slowly. We drove for a time along the heights above the river in the darkness. Long after dark, among all the cottonwoods of the river bottom, the shrieking of the cicadas rose in a wail that was almost a howl.

During the night the naked hills cooled as swiftly as they had become heated under the sun. The thermometer dropped a full thirty degrees. It stood at seventy-five when we arose at dawn. Fifteen miles south of Glendive we came upon our first lark bunting, a male in brilliant plumage of contrasting black and white. Then suddenly lark buntings were all along the roadside fences, often in pairs, the black and the brown together. We appeared to have crossed some invisible boundary line in their local distribution.

The yellow hills, rolling down to the river around us, were gashed by deep gullies, gouged out by the runoff of storms. It was such erosion that brought to light, in this region, the fossil bones of many prehistoric monsters. In Montana, scientists have found one of the world's richest hunting grounds for prehistoric big game. It was in the very region we were traversing that Dr. Barnum Brown, of the American Museum of Natural History, unearthed a Triceratops nearly thirty feet long. And farther to the west in the same state he discovered the first virtually complete skeleton of *Tyrannosaurus rex*, that giant among the "terrible lizards" of the past.

Compared to the immense stretches of time separating us from the dinosaurs, less than a century, a mere tick of the clock, has passed since this land along the Yellowstone supported its vast herds of bison. One of the last big buffalo hunts in America occurred in the 1880's in the rangeland between Glendive and Miles City. Forty years of unparalleled slaughter had stripped the plains of the great animals. Once so numerous that at one point on the western prairie a stampede took two days to pass a pioneer outpost, the wild bison are now entirely gone. Even the buffalo grass that nourished them, through the plowing of land and overgrazing by livestock, has almost disappeared.

The red man, who had hunted the bison without reducing its numbers, used nearly all parts of the animal. He tanned the hide. He dried the meat into pemmican. He even cracked the bones and saved the marrow. But the white men who followed engaged in a frenzy of slaughter that for its pitiless magnitude is equaled only by the destruction of the passenger pigeon. Sportsmen, with no other end in view than the joy

BETWEEN HAY AND GRASS

of killing, fired on the bison from train windows and shot at the swimming animals from river boats. Woodhawks, the men who supplied Missouri River steamboats with wood in summer, spent their winters killing buffalo.

In a single season, hide hunters sent 1,000,000 buffalo robes to market. Because of hasty or poor tanning, it is estimated that for every hide that reached its destination in good condition two or three animals were slain on the western prairies. Professional hunters, operating from rail lines for the export trade, would slaughter thirty or forty bison in a single savage attack; then, stripping off the hides, they would leave the huge carcasses to decay on the plains. Vying with them were the tongue hunters who killed only to obtain the tongues which were dried in the sun, barreled and shipped as a delicacy to eastern cities. Even when the meat was fully utilized, as in the famous Red River hunts of the 1860's, vast numbers of the buffaloes were slain. As many as 1,200 carts, each with a 1,000-pound capacity, followed in the wake of the hunters to bring home the spoils of the expedition.

Nobody knows the total number of bison slaughtered. It is estimated that in the eighteen years between 1866 and 1884, 12,000,000 of the great animals were shot down. Tongue hunter and hide hunter, woodhawk and sportsman turned great stretches of the prairie first red with blood and then white with bleaching bones. How enormous was their work of destruction is indicated by the fact that in later years when buffalo bones were shipped east to fertilizer plants and carbon factories, they were sometimes stacked along railroad sidings like cordwood in piles eight or ten feet high and half a mile long. In one three-year period, the Atchison, Topeka and Santa Fe Railroad transported 1,350,000 pounds of these bones, mainly from the plains of western Kansas.

Gone now are the whitened bones and the bleached skulls —those "Mormon signposts" on which the westward-traveling pioneers used to scrawl messages for those who followed. By

1890, the wild bison, which once had roamed the west in millions all the way from Texas to Saskatchewan, was making a last stand in remote wilderness areas. In less than four decades it had been pressed close to extinction. Only the efforts of alarmed conservationists saved the remnants of the great herds so that today, mainly within national and state preserves, it still survives as a species in its native land.

Below Fallon, onetime home of Berny Kempton, the rodeo champion famed for lassoing kangaroos in Australia, we came to the mouth of Powder River. Fine black sand in its bed, reminding pioneers of gunpowder, had given it its name. As it emptied into the Yellowstone, it was ending its long northward flow through Wyoming and Montana. We gazed at this stream with a somber, almost morbid, fascination. For, a century before, over the hills that bordered its banks, there had advanced a bizarre assemblage. One hundred and twelve spirited horses, forty servants, a dozen specially trained hunting dogs, an arsenal of guns and ammunition, ponderous carts pulled by oxen and filled with every imaginable luxury, all these had moved in a long procession beside the Powder River. It formed the most elaborate and the most senselessly destructive hunting party in the history of the Far West.

Across the sea from Sligo, Ireland, had come the nobleman, Sir George Gore. With his retinue of servants he wintered at Fort Laramie. Then, in the spring of 1855, with Jim Bridger as guide, his caravan started north, following the Powder River to its mouth. Everywhere the party went it laid waste the wildlife. In one protracted hunt, forty grizzlies were slain. Along the Yellowstone 2,500 bison were slaughtered, most of them left lying where they fell. So great was the meaningless bloodshed that the Indians complained bitterly to the government agents who, in turn, relayed their objections to Washington. In what was at that time a rare act of interceding in behalf of wildlife, the Federal government placed restrictions on the party and, after two years of ranging over the prairies and among the mountains like a bloodthirsty weasel in a henyard, the noble huntsman sailed for home.

At Miles City we parted company with the Yellowstone and angled down the long, dry road that runs for 175 miles to the southeast, across the tip of Wyoming and on into South Dakota. At Broadus, with cicadas loud in the cottonwoods, we crossed Powder River again, here more than 100 miles upstream from its mouth. Through the country beyond, the air was rich with the smell of the sage. At long intervals, small villages materialized beside the road, each drowsing in the heat of the day under its coverlet of dust.

Toward mid-afternoon, we pulled up beside a dry gulch surrounded by sagebrush slopes. Sparse cottonwoods shaded the coulee that opened below us. We had been there for several minutes before we became aware of two gray birds merging with the gray bark of a fallen cottonwood. Resting quietly on the trunk, they appeared almost as large as turkeys. We scrutinized them through our fieldglasses. They showed little alarm. Then deliberately, one after the other, they jumped down and walked leisurely away up the ravine where they were joined by four others. Black belly patches made positive our identification. We were seeing for the first time in our lives the famous "cock of the plains," the sage hen that Lewis and Clark discovered near the headwaters of the Missouri. It is the largest member of the grouse family. Males sometimes weigh as much as eight pounds.

Only where the sagebrush, *Artemisia,* grows are these birds found. The lives of this plant and these birds are inseparably linked together. Leaves and buds of the sage comprise the bulk of the grouse's diet. For nutritive value, scientists have found, this fare ranks among the richest of vegetable foods. The birds also turn to sagebrush for their shelter and hiding places. The gray of the bird and the gray of the vegetation are almost identical. As the little procession moved single-file out of the coulee and onto the slope beyond, we observed the perfection of their camouflage. One by one they melted into the sagebrush, visible one moment, dissolving into their surroundings the next.

All down the road before us that day, advancing as we advanced, were the heat-produced mirages of pools of water on the pavement. They accompanied us wherever we went in our summer travels, these optical illusions reflecting the clear sky in the distance and created by the juxtaposition of air of different temperatures just above the surface of the road. We saw them mostly along the asphalt and concrete highways but, on occasion, even above the hard-packed dirt of a side road. Not far from Alzada and the Little Missouri River, we met a bright-red truck that came racing down the road toward us, its image reflected and elongated on the surface of the mirage, its colors mirrored by a pool of water that was not there.

It brought to mind another aspect of color that had occupied us during earlier dawns of this journey into summer. Did we dream in color? Psychologists say some people do, some do not. Did we? Neither Nellie nor I could be positive. We thought we did; but we were not sure. So, for a week or more, each morning we tried to recall our dreams in a search for color. One dawn, Nellie reported dreaming of a bright-red warbler with delicate shadings, a bird no one had ever seen before. A few days afterwards I recalled a more prosaic dream, seeing an automobile that was blue and white, just like ours. Thus assured on dream colors, we had continued through the myriad actual hues and shadings of summer.

As we left Montana, the Bear Lodge Mountains of northeastern Wyoming, pale blue and far off, rose on our right. Ahead, to the east, the Black Hills of western South Dakota formed a dark line on the horizon. They both emphasized our dilemma.

In the range country of Montana there sometimes comes a time in spring when the winter hay is gone and the new grass has not arrived. In this period between hay and grass the cattle are hard-pressed for food. Similarly a man in difficulties is described as being "between hay and grass." During all our Montana miles we felt that we, too, in a way, were between hay and grass. We were undecided, in a quandary, pulled in several directions, uncertain which way we ought to turn next in our summer wanderings. Should we swing back across the plains? Should we turn toward the mountains? Should we visit the prairie dogs? Or should we follow the Platte River?

Before we crossed into South Dakota and reached Spearfish late in the day, we had come to a decision. The prairie dogs won. Thus relieved in our minds, we were soon relieved in our bodies as well. A thunderstorm built up at sunset, breaking the heat wave. It brought moisture and coolness. And, at its end, it lifted into the blue-black sky above the frowning heights of the Black Hills twin rainbows of unusual brilliance that remained, one above the other, for a long time before they faded gradually away.

NINETEEN

HOME OF THE PRAIRIE DOGS

THE Indians called him Wishtonwish. Lewis and Clark called him the barking squirrel. The French trappers dubbed him the little dog. Early travelers referred to him as the Louisiana marmot. Zoologists bestowed on him a scientific name meaning dog-mouse. But to millions of Americans, past and present, he is the prairie dog.

In the sunshine of that summer morning, our first prairie-dog town—with all its life and movement, excitement and humor—spread away around us. For thirty acres or more the sparse gray-yellow grass was dotted with mounds of earth. Hundreds of alert yellow-brown animals, each about the size of a half-grown woodchuck, sat up beside their burrows. They watched us just as other prairie dogs had watched the oxen and the covered wagons plodding west, as they had watched the pony express riders go galloping by, as they had watched the first longhorns replace the buffalo.

We stood that morning in a fragment of the Old West, now almost entirely gone. Once such mounds of earth spread across hundreds of millions of acres. A single dog-town on the Staked Plains of upper Texas extended for 250 miles without a break. It occupied 25,000 square miles and is computed to have had 400,000,000 inhabitants. Today, the only remaining evidence of this vast mammal metropolis is a name on the map: the Prairie Dog Town Fork of the Red River in northern Texas. Like the bison, the prairie dog is a symbol of the

HOME OF THE PRAIRIE DOGS

pioneer West. And like the bison, it has been wiped out over most of its former range and forced back into last-ditch stands within national parks and similar sanctuaries.

Almost the first colony to obtain such protection was the one that spread around us now. On September 24, 1906, by presidential decree, Theodore Roosevelt established the earliest national monument at Devils Tower, in the northeastern corner of Wyoming. For more than fifty years, within reach of the shadow of this volcanic monolith—formed like a titanic tree stump and rearing 865 feet into the air—the colony had flourished. The animals had become accustomed to people. Under the sun, on this summer day, the events of the busy town went on as though only prairie dogs were present.

Wandering about this animated scene or sitting quietly at the top of the long, gentle slope occupied by the mounds, we watched, close at hand or through our field glasses, all the comings and goings of the prairie dogs. In size they might have been cottontail rabbits, without the rabbit ears. Instead their ears were small, rounded, close-set, giving them a flat-headed appearance. Their eyes were placed well forward so they were almost the first things to appear above the surface of the ground when an animal emerged from its hole. Plump and short-legged, all the individuals of this animal town had the disarming appearance of friendly little yellow puppy-dogs with black-tipped tails. These stubby tails they wagged vigorously, not from side to side but up and down. It is this vertical flipping motion that reveals their kinship with their smaller cousins, the ground squirrels.

Scores of the prairie dogs stretched out on the ground while companions carefully went over their fur, cleaning and dressing it, sometimes with their teeth pulling out the loose pelt for a surprising distance. A few wriggled and squirmed and rolled, taking dust baths to rid themselves of fleas and ticks. Others, mainly the smaller, younger animals, seemed engaged in games. With a great racing through the grass, one would pursue another until, like a football tackler, he bowled him over and sent him tumbling. Then their roles would be reversed; the chaser became the chased. Although their gait appears short-legged and floundering, they covered distance with unexpected speed. About these pursuits, as about almost all the activity of the slope, there was a playful, good-natured quality.

Few other animals are so social and amicable. All that day we watched prairie dogs go visiting. Whenever they met in their foraging they paused in greeting, touching noses as though bestowing a kiss, fondling and patting one another, often sitting for some time together with the forepaws of one resting on the shoulders of the other. Once we saw two sit up facing each other, place their forepaws together with repeated pattycake motions, then drop to the ground and touch noses.

How many prairie dogs did we see that day? How many were living in the subterranean city around us? Nobody knows. They kept coming and going, appearing and disappearing. At any given moment, what proportion of the population is above ground and what proportion is out of sight below it? Who can tell? An exact census of a dog-town is virtually impossible. The nearest approach park officials could make, they told me, indicated the figure was somewhere in the neighborhood of 3,000.

It was pleasant sitting there in the morning sunshine and for a time we closed our eyes and listened to the sounds of the animal city that came to us from every point of the compass. There were yipping notes, a churring sound, and sharp, staccato chips. The so-called bark of the prairie dog is largely a magnified squeak. Sometimes it resembles the yelp of a small puppy. But most of the time it suggests the sound of a squeezed rubber doll. A century ago, John James Audubon noted the close similarity between certain calls of the prairie dog and those of the western kingbird.

At times a yipping animal will bounce up and down as though it is on springs. When it gives its loudest, most far-carrying call, the prairie dog stands erect, lifts its forelegs as in a salaam, and throws itself upward and backward as if propelling the sound from its throat. Young animals sometimes jerk so violently they lose their balance and throw themselves over backward. This call has become recognized as a territorial bark. It broadcasts prior claims to the burrow and the ground surrounding it. Usually it is answered from other mounds nearby. On numerous occasions we saw a chain reaction of such calling run across the town, animals all down the slope rising on their hind legs, pointing their noses at the sky, throwing up their forepaws, barking their loudest, their lighter underpelts catching the sun.

One call—a repeated birdlike "Chip! Chip! Chip!"—appeared to be a note of alertness rather than of alarm. Usually when we approached a burrow, its owner would take a position lying across the opening of the hole, ready for an instant plunge to safety. There it would repeat this sound, its black-tipped puppy tail flipping up and down with every "Chip!" Every minute or so, when a prairie dog is feeding away from its burrow, it sits up and looks intently around. At the first suggestion of danger, it raises a shrill alarm. This is echoed from end to end across the colony. All the animals sit up, watchful, ready to whisk from sight. In this way a constant surveillance is maintained of the sky above and the country around. The householders of a dog-town are continually kept informed by scores and hundreds of town criers calling the news.

So effective is this warning system that rarely are prairie dogs caught in the open even by such aerial enemies as golden eagles, rough-legged and red-tailed hawks. Sometimes a prairie falcon will fold up in the sky and thunderbolt down on some animal that has wandered too far away from its burrow. But such occasions are rare. One pair of falcons nested for several years high on the side of Devils Tower. Although they hunted over the area daily, they rarely wasted time stooping on the prairie dogs. How alert these animals are to danger from the air we observed when a considerable portion of the colony disappeared when one of the larger songbirds flew overhead. It is the extremely rare black-footed ferret, racing through the maze of tunnels, and the badger, able with its shovel forefeet to dig out a burrow in as short a time as fifteen minutes, that are the arch-foes of the prairie dog. Hereditary enemies such as the coyote, the fox and the wolf are usually detected long before they reach the outskirts of the city.

Most of the time the prairie dogs we saw reappeared from their burrows within a few minutes after they disappeared. A serious fright, however, will keep them underground for an hour or more. In the brains of these little mammals their burrows and safety are synonymous. Even when shot, a prairie dog will often tumble into its hole and kick itself downward in its death struggles. This led the English traveler, John Bradbury, who visited the region of the Dakotas in 1811, to report that the creatures were so quick that after they saw the flash of the gun they could dive to safety before the bullet reached them.

For a long time we watched one large prairie dog in a puzzling performance. It was working around the entrance of its burrow, patting down the earth, when first we noticed it. Then, placing its forepaws wide apart on the ground and ducking its head, it lifted its hindquarters until it was on tiptoe. It suggested a man about to stand on his hands. In this position, it lowered its head vertically and pressed its broad nose into the dirt. Over and over again we saw it repeat this odd action. The forepart of its body lifted and fell like a pile driver. Then it would shift to a new position and repeat the process. What was it doing? It appeared to us to be tamping down the earth with its nose. And that was precisely what it was doing. The mounds of dirt that mark the burrow entrances are in reality dikes. Ringing the holes and rising to a height of as much as two feet, they not only provide the animals with watchtowers but prevent the water of summer cloudbursts from pouring into the tunnels. The rarer, less social white-tailed prairie dogs of the higher mountain valleys —where drainage is better—merely pile the excavated earth beside the hole. But the black-tails of the plains carefully shape it into a protecting wall, thoroughly tamping the dirt into place. When we examined the craterlike opening where the large prairie dog had been working, we found it stippled all over with the indented prints of its nose.

The crater form of the burrow entrance has a special value for the prairie dog. He can hurl himself into it from any angle. You often see little clouds of dust puff upward as the animals strike the sides in pouring themselves at high speed down these funnels of earth. The burrow that lies below is far from a simple hole in the ground. It is, in fact, one of the most elaborate dwellings prepared by any American mammal. It begins with a plunge hole that descends almost vertically. A yard or so below the surface, a niche or shelf is cut in the side of this tunnel to form a "guard room" or listening post. Here the owner pauses when he enters and when he leaves. And here, in safety, he barks defiance to his enemies outside. Lower down, the tunnel branches into connecting passageways with spare rooms and a round sleeping chamber lined with dry grass. Not infrequently a corridor ascends straight up from the main tunnel for several feet, ending in a vacant room. This is believed to be an escape chamber in case of flooding, the air compressed upward into such a space keeping the occupant alive until the water drains away.

In the early days, during the building of one of the transcontinental railroads, a steam shovel cut straight through a prairie-dog town, laying bare the labyrinth of tunnels that descended for as far as fourteen feet and extended horizontally for as far as forty feet. In the course of their extensive burrowing, the animals plow up the soil like oversized earthworms, making it more porous and breaking up the hard, sun-baked crust of the prairie. Prairie dogs literally turn the land upside down. For they not only bring up soil from below, but their abandoned burrows gradually fill with the dust of the surface soil which, in this manner, is shifted to a lower level.

Some years ago, when Dr. James Thorp was regional director of the U.S. Department of Agriculture's Soil Survey for the Great Plains, he and a fellow scientist made some interesting calculations on the amount of earth prairie dogs bring to the surface. They measured off a four-acre plot in a dog-town on the high plains of eastern Colorado. In that relatively restricted area, they found the amount of sand, gravel and subsoil excavated by the animals totaled 311,950 pounds—156 tons for the four acres, thirty-nine tons to the acre. Here the average number of holes was seventeen to the acre. Near Carlsbad, in New Mexico, Vernon Bailey encountered 1,009 holes in a twenty-acre tract, an average of fifty to the acre. It was the opinion of C. Hart Merriam, Chief of the U.S. Biological Survey, that a conservative estimate of the number of holes in all prairie-dog towns would give an average of twenty-five to the acre. The immense churning and rejuvenating effect on the soil resulting from the activity of these animals must have been an important factor in the health of the prairies.

The most tenacious misconception in regard to the tunnels of a dog-town is that indestructible fable that rattlesnakes, burrowing owls and prairie dogs all live amicably together in the same hole. In the main, the owls and snakes occupy deserted burrows. When the opportunity offers, they both kill and eat small prairie dogs. A more careful reading of the first account of the prairie dog, in the journals of Lewis and Clark, would have exposed the falsity of the fable in the beginning. On September 7, 1804, members of the expedition killed a rattler in the first dog-town they encountered on the banks of the Missouri River. They cut it open and found a young prairie dog inside.

HOME OF THE PRAIRIE DOGS

Here at Devils Tower, as well as later at Wind Cave National Park, in the Black Hills of South Dakota, and at the celebrated walled-in city of the prairie dogs at Lubbock, Texas, we watched earth being brought to the surface. Once we were no more than fifteen feet from the entrance to the hole. The owner would appear at intervals, sometimes pushing a load of earth before it with its strong forepaws, more often emerging backward and kicking the dirt out with lusty back-sweeps of its hind feet. During an interval of quiet, one of the younger animals, endlessly curious, came running up and peered into the hole just in time to be met with a barrage of flying earth that sent it tumbling over backward in surprise. Occasionally a digging prairie dog will stop work to bite out a root and eat it as a tidbit. Before his burrow is completed, he may bring as many as twenty bushels of excavated material to the surface.

Toward noon that day, as on every summer day, the activity of the prairie dogs quieted down. More and more they drifted out of sight, retiring underground for their midday siesta. It is early in the morning and late in the afternoon that their activity is greatest. A heavy shower will always send them indoors. They are creatures of the sun, and summer is their season. During the winter, although they do not hibernate, strictly speaking, and may be seen abroad for short periods on bright days when the temperature stands at zero, they drowse away most of the months of cold. Largely they live on themselves, using up the reserves of fat stored up late in summer. Already we saw some so layered with fat that when they turned to look at us over their shoulders their sides crinkled like corrugated paper.

While the prairie dogs napped in their burrows, we ate lunch at a small restaurant at the edge of the national monument. Here little bags of dry bread were piled up under a sign reading: "Prairie Dog Food." Normally the food supply on the Devils Tower slope would be insufficient to support so large a colony. But here the regular seed and vegetable fare is supplemented by handouts from tourists. And a fearful and wonderful assortment is included. Bread and cookies, marshmallows and popcorn, peanuts and crackers, cake and candy, even bubble gum is welcomed by the voracious prairie dogs. I saw one chewing away on a cellophane bag that had contained potato chips. So far as I could see it was devouring the cellophane with relish. Another ran its forepaw around inside the neck of an empty Pepsi-Cola bottle and licked off the sweet residue. We were told of one prairie dog that kept returning to eat more ginger snaps, although it scolded and chattered and struck its nose with its forepaws each time its tongue began to burn. As we gazed over the almost deserted slope, we wondered how many midday siestas were being disturbed by stomach-aches.

That afternoon, besides distributing dry bread—and the drier it is the better the prairie dogs like it—we provided what we thought was a new taste sensation for dog-town. We passed out half a dozen fig newtons. Left to themselves, the animals feed on a wide variety of grasses and herbaceous plants. We would see prairie dogs nip off grass stems and run the heads rapidly between their teeth, as through a machine, to strip away the seeds. Frequently they harvested a tiny form of milkweed that grows to a height of only a couple of inches between the burrows. Another plant they seemed to favor at Devils Tower is a miniature mallow with the picturesque name of cowboy's delight.

In winter, when other food is scarce, prairie dogs will sometimes nibble between the spines at the flat, fleshy leaves of the prickly-pear cactus. On several occasions in the Southwest, during great grasshopper outbreaks, the animals have been seen turning from their vegetarian diet to feed voraciously on the insects, rushing this way and that, pursuing and pouncing on their active prey, sometimes even leaping into the air in an effort to catch grasshoppers on the wing.

About 75 per cent of the plants consumed by prairie dogs

HOME OF THE PRAIRIE DOGS

are considered of value for grazing. It has been calculated that 256 of the social animals will eat as much as a cow, and thirty-two will consume as much forage as a sheep. Without doubt, extensive dog-towns and ranching are incompatible. Consequently every stockman's hand has been against the "little dogs" since early days. The government joined in. Tons of strychnine were dispensed free to western ranchers. Year by year it took its toll. Cyanide and poison gas added to the destruction. In Washington, officials announced triumphantly, during the 1920's, that colonies were being wiped out at a cost of seventeen cents an acre.

This war of extermination went on until on the Great Plains today the prairie dog is virtually as rare as the bison. Across vast tracts that had been filled with sound and movement and exuberant life a great stillness settled. Wind-blown dust filled and obliterated the ghost towns of the tunnels. Only in protected sanctuaries were these marked animals safe from persecution. In all probability it will be these sanctuaries alone that will save the species from extinction. This achievement is something to rejoice over. For no living species should be pushed to complete extinction—least of all one that so obviously enjoys its life and lives it to the full as does the friendly, sociable little prairie dog.

In 1924, at the height of the poison campaign, the Department of Agriculture issued a bulletin on damage done by prairie dogs. Without examining a single stomach to discover all that the animals were eating, the authors concluded: "So far as these tests do indicate the prairie dog does not possess a single beneficial food habit." The "so far" of the experiments, it now has become obvious, was not far enough. In 1939, a biologist with the same department, Leon H. Kelso, reported on exhaustive studies in which he examined the stomachs of more than 500 prairie dogs. His researches revealed beneficial habits unmentioned before. Some of the animals had eaten Russian thistle and rabbitbrush, both unsuitable for cattle. One stomach contained 20,000 seeds of the noxious knotweed. Others, in Montana, showed that as high as 70 per cent of the food consumed represented loco weeds, poisonous to livestock. Fourteen out of twenty stomachs collected in May contained cutworms. They represented as much as 35 per cent of the food eaten. A single prairie dog was found to have consumed, and thus destroyed, 600 of the injurious larvae. The 1924 report, giving the stockmen what they wanted to hear, had played its part in the relentless war on the prairie dogs. By the time the carefully prepared 1939 report appeared, it was too late to help. The slaughter of the unprotected animals had been carried almost to completion.

Toward sunset that afternoon, we sat near one of the mounds and watched its owner foraging for food. It is only on rare occasions that a prairie dog goes as far as 100 yards from its burrow. In this instance, the farthest foray of the animal carried it no more than thirty feet from its doorstep. Not infrequently it would gather a large mouthful of vegetation and then return to dine at leisure beside the entrance, both forepaws held to its mouth, its bright eyes peering over them, alert and watchful. Around every burrow the ground is almost bare, and all across the length and breadth of the colony the view is kept unobstructed. Heavy plants are almost invariably clipped off close to the ground soon after they reach a height of six inches.

As a result of observations made in two burlap-enclosed towers during three summers and one winter at the Wind Cave National Park, in South Dakota, John A. King discovered that a prairie-dog colony is really broken up into smaller units or coteries. Writing in the October, 1959 *Scientific American*, he reported that each unit is made up of a number of individuals or families and that their common territory has definite boundaries which they jealously defend. This division of the town into smaller territories, he believes, aids in the distribution of the population and reduces the chances of overgrazing.

The prairie dog we watched foraging in the sunset was obtaining both food and drink. Most of these animals, when living in a wild state, may never taste water in their lives. They obtain a certain amount of moisture from the vegetation they eat. But even this is not necessary. For, like pocket mice and kangaroo rats and other rodents of the dry country, their systems are able to turn the starch of dry seeds into water. One scientist fed a pet prairie dog perfectly dry food for a week and a half. It received no liquid of any kind. At the end of the period, he offered it a dish of water. It took only a perfunctory sip, no more than when it had water available all the time.

Early settlers in the West refused to believe that prairie dogs could go indefinitely without water. For decades the mistaken idea was current that the inhabitants of every dog-town sank one shaft straight down to reach the flow of subterranean fountains. So convinced of this were many pioneers that they took special pains to dig their wells where prairie dogs had lived. They were baffled by the dry holes that resulted. In one instance pipes were driven to the depth of 1,000 feet without striking water.

With the going down of the sun that evening, the shadow of the towering monolith spread farther and farther across the slope and over the mounds that dotted it. To us it seemed symbolic of the protection this natural wonder had extended to the prairie dogs close by. In the shadow we saw the animals sitting up beside their burrows or visiting their neighbors for a last time or seeking some special tidbit to finish the day. Their calls became fewer; their activity lessened. Steadily their numbers decreased as more and more dropped away to sleeping quarters underground. By eight o'clock hardly a prairie dog was in sight. Quiet lay over the dog-town. But it was the quiet of rest, not the stillness of death.

When we left, after the dusk had deepened, all these peaceful burrowers, so vulnerable to man's attack in their close-packed, social communities, slept curled in their nests of grass just as they had slept in that long-gone time when the Great Plains were called the Prairies of Louisiana. The morning sun would bring them to the surface again. It would renew their vivacity, their hearty spirits, their transparent enjoyment of life. For here prairie dogs, in this far corner of their once-great domain, had found a permanent sanctuary.

TWENTY

DRY RAIN

SUNDANCE dropped away behind us. In the cool of that dawn we skimmed over the rolling land like a swallow flying. We were in a slanting world. Everywhere we were going up or going down with no extended stretches of level ground between. On each successive crest our eyes' reach extended swiftly like an outstretched arm. We crossed dry creeks. We ran by snow-white patches of gypsum cropping out of the bleached yellow of the rangeland. At long intervals, coneflowers waved beside the road resembling purple daisies grown abnormally large. Past blooming now, the yuccas supported on upthrust stems the pale-green ovals of their massed seed pods. Meadowlarks, all along the way, sunned themselves on fenceposts. Nearly all were silent. For this was August. The time of the singing of birds was mainly over.

Riding through these surroundings of ever-changing interest, I found my mind wandering back among the crossroads, the shifts in plans, the innumerable factors of chance and design that play a part in bringing us to any one point in space at any one moment in time. Each such arrival contributes its grain of weight to the sum total of our experience. Other grains would have made other lives. These were the ingredients of ours. On this earth there are, in truth, so many new experiences that I desire but rarely to live over any part of my past life, not even the happiest moments of the finest hours I have known.

Long before mid-morning we had left behind the Wyoming rangeland and had crossed into South Dakota. Once more we were among ponderosa pines, the conifers that make the Black Hills black. We swung southward, traversing the length of beautiful Spearfish Canyon. Near Eleventh Hour Gulch we stopped to watch white-throated swifts whirl above the cliff tops and violet-green swallows flutter, alight and cling to cranny edges as they fed their young. Down the whole twenty-mile length of the gorge, down a varying ribbon of wildflowers —pale purple asters, yellow tansy, lavender bee balm—the butterflies wandered. For one memorable moment around us the chasm rang with the sweet cascading song of the canyon wren. And before the face of each varicolored cliff, grasshoppers danced. The walls of rock behind them, like giant sounding boards, magnified the fluttering crackle of each insect serenade.

Beyond the southern mouth of the canyon, through hills and open land, in forest and over sparsely settled country we ran on, hour after hour. Once on a lonely stretch of mountain road we passed the body of some dead animal surrounded by a dense, dark cloud of flesh flies. I marveled then, as I had marveled years before in a similar moment amid the Great Smokies of Tennessee, at the speed with which these insect scavengers gather from afar to deposit their eggs.

That night we settled down in Hot Springs, not far from the Wind Cave National Park and its prairie dogs. Then we were on the road again, riding once more toward the south. This time we were bound for a new world, the world of the 20,000 square miles of sandhills that extended away beyond the Nebraska line.

All life, under the burning sun of mid-morning, seemed lying low, keeping silent, all except the cicadas. Intoxicated by the heat, they shrilled from the sagebrush of every open slope along the way. As we went south we met the wind head-on, a dry wind that whipped the feathery sage and sent it, with the hard drive of its gusts, streaming and waving like masses of ferns. The air was hazy with heat and dust. The distant hills became more dim and vague in outline. Not far from the Nebraska line we topped a rise and found a wide valley of farming land spreading away before us. All across it what looked like clouds of yellow smoke were pouring upward from the open ground. Sweeping across the fields, swiftly drying the top of plowed or harrowed ground, the arid gale was lifting the surface of the land into the sky as dust. And beyond the valley, for a hundred miles and more to the south, all the open country was adding other billowing clouds to the tons and acres of air-borne soil. We drove on and entered the dry rain of the forward fringe of our first big dust storm.

According to the United States Department of Agriculture, during the year of our summer journey 3,864,000 acres of the Great Plains were damaged by blowing. Although such storms may occur at all times of the year, they have been most destructive in early spring. In the dust-bowl areas of Nebraska, Kansas, Oklahoma, Texas, New Mexico, Colorado, Wyoming, Montana and the Dakotas, there are millions of other acres where, given the right conditions, unprotected surface soil is ready to be carried away by the wind. The long sequence of dust storms during the drought years of the early and middle thirties turned 18,000,000 acres of farming land into yellow-hued desert. In the midst of this disaster, that staple of early days that helped American pioneers endure the hardships of the frontier—the humor of exaggeration—came to the fore again. There were stories of areas so dry the cows gave powdered milk, of prairie dogs six feet in the air trying to dig their way down to the earth, of the man who poked his finger into the air and could see the spot for a week afterwards.

The quantity of soil removed in a given dust storm depends largely on the dryness of the ground and the speed of the wind. And these factors, in themselves, are interrelated. The wind-borne soil, on this day, rushed past us on a forty-mile-an-hour gale. A wind moving at this speed, scientific experiments have revealed, has four times the drying power it has at twenty miles an hour and sixteen times the power it possesses at ten miles an hour. The drier the air, of course, the more rapidly it sucks moisture from the ground. But in the case of the same wind, the drying effect on the surface of the soil increases as the square of its speed.

The swiftness of this drying process at times may produce astonishing sights. Later in our trip we observed a dust devil spinning down the length of a sand bar almost awash in the Cimarron River. In one Nebraska stream, a friend of ours, Dr. James Thorp, head of the geology department at Earlham College, in Indiana, once saw this process of surface drying accomplished with such dispatch that dust was blowing away from the top of a wet mudbank.

As we drew near the Nebraska line, the sunshine dimmed to an odd, filtered, yellowish light. The sky took on a sickly cast of yellow-gray. On every side the hills became smoky and indistinct in the wind-blown dust. Hollows, dimly seen from the crests of the rises, appeared filled with a dense, yellow-tinted fog. The dry smell of dust now was strong in the air.

Out of this murk we saw combines careening down the road. They were trying to outrun the storm, heading north toward that zebra landscape of alternating strips of gold and green, of grain and fallow land, that we could picture so well. As we continued south the rim of the prairie faded. The walls of dust pressed closer. We advanced into a steadily contracting world, meeting head-on the wind that, ever more heavily laden, came sweeping out of the south.

All across the North Dakota prairies we had seen windbreaks planted to the north or northwest of the farm buildings. In parts of Nebraska we were to see them planted to the south. One day we passed a whole grove of cottonwoods, trees forty feet high, that recorded the prevailing direction of the storm winds. Every trunk, like a partly-drawn bow, curved

NELLIE searching for summer birds along a trail that follows the overgrown summit of a ridge in Minnesota.

RUSHES of the Souris refuge, above. Below left, a dry stream bed in Montana. Right, a gully cut by erosion.

toward the north. That the winds of this area so often blow from the direction of the hot, dry land to the south is undoubtedly a factor in building up the dust storms of the region.

We peered into the billion-particled clouds around us. We watched the tiny, uncountable fragments of the earth sweeping past us on the wind. Where had they all come from? How far had they traveled? Some were at the very beginning of their journey. We saw new clouds of dust peeling from the length of every roadside field. And we, ourselves, added a cloud of our own when we came to a mile of construction where, looming up suddenly before us, heavy grading equipment suggested the prehistoric monsters that once roamed this land. But other components of the storm no doubt had come from farther to the south. We wondered how much of the dust around us had had its origin in that famed graveyard of the dinosaurs along the upper reaches of the Niobrara River. We wondered how much had come from those eroding bluffs where the discovery of the *Daimonelix*, the Devil's corkscrew—now believed to be the fossilized remains of the spiral burrow of some extinct rodent—set off half a century of scientific debate.

Over the instrument panel before us a fine yellowish dust had formed. I ran a forefinger across it. It had no gritty feel. The powdery material was smooth like flour. This fine dust may well have been carried northward by the wind from the famed loess regions of southern Nebraska. Formed of pulverized rock and deposited by the wind, loess, the "golden earth" of agriculture, forms the richest grain-producing areas of the world. It supports the densest populations on earth, the teeming millions of China and India. The precious yellow-brown loess of China, which colors the Yellow River and the Yellow Sea, has a depth of as much as 600 feet. Of German origin, the word means loose or open-textured. In farming land of the Midwest loess combines fertility with a capacity for absorbing rainwater like a sponge. "Golden earth" is to

Nebraska what "black gold" is to the oil regions. It represents the agricultural wealth of the state.

Looking out into the swirling dust clouds, we could imagine ourselves in the very midst of the creation of these fertile lands. For it was in the wake of the grinding northward retreat of the glaciers that winds, winds that ceased blowing millions of years ago, swept the powdered rock aloft and deposited it over the countryside. Loess has been found on the flanks of mountains 12,000 feet above the sea. Nebraska's most fertile land is derived from these dust storms of long ago, from minute air-borne particles, ground in the mills of the glaciers and deposited during the Pleistocene.

One August noon, later in the trip, we stopped for lunch at the Lakeway Hotel, at Meade, in southwestern Kansas. Before I got into the car again I looked up at the third-story windows. One of those rooms had been occupied in the late-summer of 1933 by a scientist, John C. Frye. At the end of a two-hour dust storm, he carefully collected samples of the material that had settled to a depth of a sixteenth of an inch on enameled surfaces in the room. Later laboratory examination of this aeolian dust revealed that almost all the particles were approximately the same size. They ranged in diameter from $\frac{1}{100}$ to $\frac{5}{100}$ of a millimeter. The wind sorts the material as it transports it. The smaller the fragment, the farther it goes. Coarse sand, for example, not over one millimeter in diameter, is usually carried no more than a mile. Fine sand may be transported several miles. Coarse silt, $\frac{1}{16}$ to $\frac{1}{32}$ of a millimeter in diameter, may travel 200 miles. Medium silt, $\frac{1}{32}$ to $\frac{1}{64}$ of a millimeter in diameter, may be wind-borne for 1,000 miles. And the finest silt, less than $\frac{1}{60}$ of a millimeter in diameter, scientists have found, may travel around the world. This difference in the distances traveled by particles of varying size accounts for the uniform character of any given area where loess is deposited.

The finer material not only travels farther but ascends

higher into the sky and requires less violent winds to transport it. A study of the loess of Nebraska, which ranges in color from nearly white through yellow and olive to very pale brown, has shown that the size of the particles becomes steadily smaller as the distance increases from the original source of supply. But whatever their size, the rock-flour fragments are laid down in countless billions. One single modern dust storm, it has been calculated, transported 70,000,000 tons of earth through the air.

Once carried aloft, the finest fragments are able to float on a slight breeze, to ride on the last remnants of a dying gale. It is the larger particles that progressively drop from the sky as the speed of the wind slackens. Ordinarily solid sand grains rarely rise more than a few feet above the ground. It is for this reason that travelers in desert regions often come upon strangely eroded rocks, shaped like toadstools. The sandblast of the gales, concentrated close to the ground, has eroded away a large portion of the rock to a height of several feet, leaving the upper portions relatively intact. In bare, dry land the wind becomes an abrasive tool. Its effect is like that of the sandblast that cuts grime from the stone of city buildings.

In the sweep of the dry rain that swirled around, the dust represented fragments from infinitely varied rocks. Even gold and other precious metals sometimes ride through the sky during such storms. Around Custer, in South Dakota, the dusty air at times glints and sparkles with minute snowflakes of mica. It is when the tiny fragments of the harder rocks predominate that the abrasion of the wind-hurled particles becomes a serious problem. Window panes develop ground-glass exteriors. Tin is scoured from plates and cans. In some arid regions only metal telephone poles can be employed. Wooden poles are soon cut down by the erosion of the flying sand. Even steel rails, in time, may be worn thin by the sandblasting of repeated gales.

Across the rock-strewn floor of a desert this abrasive action of the wind has another significance. Here the particles produce other particles. The wind-driven sand is the principal agent in wearing away the rocks and pulverizing the fragments. While most of the rock-flour of loess was ground in the mills of the glaciers, more is being added by the action of the desert winds. Here it is the wind that creates and it is the wind that distributes that great boon to man, the fertility of loess.

It was about three in the afternoon when we entered Nebraska. The small village of Wayside, invisible behind the curtaining dust, was to our right as we crossed into the eighteenth state of the trip and turned east on our way to Chadron. From time to time a straggling car, with headlights on, groped its way toward us. In the storm, static electricity was building up. Silk crackled. Metal sparked when I touched it to the instrument panel. At the height of a great dust storm hair sometimes stands on end, barbed wire fences become charged, the ignition systems of automobiles are disrupted.

Once we were almost on a large sign when it suddenly loomed up beside the road. It notified truckers of a Nebraska "Port of Entry" that lay ahead. It struck us as a humorous note to call anything a "port" in this dry land blowing away in clouds of dust. Yet I suppose that Nebraska has more right than many states to the term. With its winding streams, it is said to rank near the top in its number of miles of river frontage. Its name, Nebraska, is an Indian word meaning "Shallow River."

As the distance to our journey's end lessened that afternoon, we took stock. We had dust in our throats, dust on our tongues, dust in our noses. Nellie was sure she could feel dust in her stomach. Our sunglasses, we found, were coated with a thin powdery layer. I discovered my eyebrows were dusty. And so we came, sometime after four o'clock, to Chadron on the northern edge of the Nebraska sandhills. We glimpsed them vaguely, indistinct and wraithlike behind billowing curtains of dust. As we saw them then they were nondescript in hue, indefinite in outline, as unreal as the hills of a dream.

Toward evening the wind slackened and the dust thinned ever so slightly. Immense and round, like a pale-gold harvest moon, the sun sank slowly. First its lower rim, then its lower half, then all became obscured behind the dense wall of airborne particles that rose above the western skyline. A delicate, golden-tinted light, new to our experience and memorably beautiful, filled all this dusty sunset. Long after the twilight had fallen, far into the night, the wind and the burden it carried continued to sweep over Chadron. But by dawn the air was almost still. Most of the dust had settled. We set out once more, with vision expanded, to enter that world of surprises, a land of green dunes.

TWENTY-ONE

GREEN DUNES

ON THIS morning I started out remembering the contrary sheep. I had met them some years ago during a night of rare insomnia. For the first time in my life I tried that proverbial rite: counting sheep jumping over bars. In my mind's eye I brought the whole flock into vivid focus. I lined up the animals facing the bars. The first sheep sailed high over the obstacle—unnecessarily high, it seemed to me. The second sheep, at the last moment, braced its feet and slid to a stop. The third tripped over the top bar and went sprawling. By that time I was more wide awake than ever, waiting to see what the next sheep would do.

I was paying, I suppose, the price for a naturalist's knowledge, for the kind of mind that sees all living creatures as individuals rather than in masses. Long since I have learned that whatever has life has individuality. No two ants, no two sparrows, no two cows, no two children are ever identical. It is only our lack of perception that leads us to conclude that all bees in a swarm, all fish in a school, all sheep in a flock are just alike. And so it is with the years and the days and the seasons. No two are ever the same. This season of our summer trip had been characterized by more than average rainfall. Everywhere we went we saw the innumerable consequences written in green.

From my boyhood in the dune country of northern Indiana, I expected to find the sandhill region of Nebraska an area of bare-topped mounds and drifting sand. Instead, as we wandered east to Valentine and south to Thedford and west to Scottsbluff, we traversed a sea of green hills that rolled away to the horizon. All the high knolls of sand had been clothed and stabilized by a surface layer of grass. The dunes were green dunes. And in the hollows between spread the luxuriant "hay meadows" that had attracted the first settlers.

For 24,000 square miles, for more than 15,000,000 acres, for an area sufficient to absorb the states of Massachusetts, Connecticut, Delaware and New Jersey, the largest area of dunes on the North American continent extended around us. Mainly in the north-central part of the state, it occupies fully one-fourth of the area of Nebraska. It is a land of hay and cattle. In pioneer times, hay was used to pave temporary roads across stretches of open sand. It was also twisted into tight rolls and burned in the stoves of homesteaders. Grass was then, as it is today, the wealth of the region. Here the signs we had seen so often along the way, "Help Prevent Forest Fires," became transformed into "Help Prevent Grass Fires." The green dunes represent the richest beef-producing region of Nebraska. They support pure-breds, Angus and Herefords, black cattle grazing across one range, red-and-white cattle across another.

But always just below the wealth of the grass lies the sand. During years of drought, every break in the sod is probed by the wind, expanded by the gales, often turned into blowouts of extensive proportions. To dwellers in the area, dunes scarred by such blowouts are "whitecaps"; the storm winds that produce them are "howlers." Even in this year of exceptional moisture, when an almost unheard-of fifteen inches of rain had fallen in the region, we saw numerous evidences of the loose, fine-grained material that forms the substance of the hills.

Where boundary fences climbed steeply up a slope, they were bordered by lines of raw, pale-yellow sand. There the hoofs of the cattle, walking in single file, had broken through the grass. Around the wooden windmills where the rangeland animals came to drink, the ground was trampled and the sand lay bare. These windmills were never high and towering; they were low, sawed-off, often no taller than a man. Once we came upon a line of curious two-toned telephone poles running across an open field. Each pole was weathered almost black except for a striking band of satiny yellow that encircled it a few feet above the ground—the product of the constant rubbing of the cattle. Around the base of each pole there ran a ring, a circular path of open sand.

All through these green dunes, houses stood remote and far apart. Side roads were few. We were in a wild, lonely, windy land with a special beauty of its own. After the fire of the sun had burned low at the end of the first day, we wandered for nearly a hundred miles through the "lake country" south of Valentine. Much of this area, embracing the greatest concentration of small bodies of water found among the dunes, lies within the boundaries of the Valentine National Wildlife Refuge.

Already, pintail and blue-winged teal, the earliest migrants, floated in rafts on the larger ponds. Wilson's phalaropes swam in clusters, spinning like water beetles. We counted more than 500 of these robin-sized shore birds on two small ponds. They all were whirling endlessly in tight little circles, pin heads lifted on pin necks, needle bills stabbing at the water as they fed. Turning about in almost exactly the same spot, one Wilson's phalarope was once seen making 247 revolutions without stopping. The feeding activity of these birds is of considerable benefit to man. Their diet at this time of year consists largely of mosquito larvae.

The solitary charm of the dune country is best appreciated in the evening and in the early morning. It was being accentuated, when we started out next morning, by the subtle shadings of the dawn light. At the time we awoke, I remember, neither of us had the slightest idea what day of the week it was. And what did it matter? For this glorious season we were living by sun time, by moon time. We had no other appointments than those with nature. We were nearing the Niobrara River, that morning, when our speedometer recorded the 10,000-mile mark of our journey into summer.

Blooms of the night, yellow primrose and white evening star, *Mantzelia nuda*, were still open in the dawn. All across one wide stretch where Queen Anne's lace bloomed, red-and-white Herefords fed as though in low-lying mist. In the rich meadows between the dunes, haying was over and new stacks rose close together like the tents of an encamping army. The sun rose; through rents in the broken eastern sky slanting rays descended.

It was in such a spotlight that we saw crossing the road before us a home-going rattlesnake. We slowed down. It coiled on the highway. I pulled up beside it. It gave no ground. We looked almost directly down on it from a distance of hardly more than a yard. Its head was lifted, weaving from side to side. Its rattles blurred and shrilled. Its coils were in constant motion. Its shifting body seemed at once fluid and as taut as a tempered steel spring. Never had I seen so lively a rattlesnake. Later that day I talked with a long-time resident of the region. He spoke with a kind of mournful pride of the superior aggressiveness of the sandhill rattlers. They were always agile and active, set on a hair trigger. They were far more dangerous, he assured me, than the sluggish snakes of the mountains.

Looking down on the flat, weaving head, on the slide of the coils, in the dawn light of that day, I gazed at this fellow-creature on earth across an unbridgeable gulf. Nature is not always as we would have her on a pleasant day. Good will toward all living creatures is not enough. Understanding that the deadly serpent was born with its fangs and venom—that it did not invent them; that the credit for that belongs elsewhere—is not enough. The mortal threat remains. The man of good will turns away from the venomous reptile with a troubled mind. For here is menace on one side and enmity on the other that go back too many years to count.

North of the Dismal River, at Thedford, we turned west to follow the road of the "potash towns" through the heart of the sandhills. In the days of the First World War, potash for gunpowder was obtained from the dry lake beds of the surrounding dunes. Factories hastily took shape. Communities along the way became boom towns in a period of short-lived prosperity. We stopped at one, about midmorning, to buy ice-cream cones. As we ate them in the heat of the day, Nellie recalled a friend of ours in Indiana who considered the finest weather of the year to be the most sweltering days of summer heat waves. He had spent his life in the ice-cream business, and blistering hours were associated pleasantly in his mind with peak sales and increased profits.

As I think back over our memories of the heart of the sandhills, two—one concerning birds, one concerning plants—return with special vividness. We had come to a little saddle at the top of a rise on a particularly lonely stretch of the road when we discovered six sharp-tailed grouse ahead of us. A seventh, apparently a young bird, had been killed by a passing car. All of its six companions seemed bewildered by the tragedy. They wandered about the crushed body or stood irresolute in the low vegetation of the roadside. None flew away when we came to a stop beside them. Even under normal conditions these birds are surprisingly tame at this time of year. To keep more of the little family group from being hit by speeding cars, I tossed the body of the dead bird up the embankment off the highway. Even then the grouse retreated only slightly. They remained nearby as though bewitched, chained to the spot, unable to leave. The huddled flock stood still in full view when we drove away.

The second recollection is associated with a wide valley among the sandhills east of Hyannis. The green wall of a ridge rose on our right as we crossed the flatland. For nearly a mile along its lower flank ran a broad, irregular band of brilliant yellow. We stopped and studied the distant stripe through our field glasses. Here, as we had never had it before, extending acre after acre, was that strange leafless plant, the naked dodder.

Love vine, pull-down, strangleweed, devil's hair, hairweed, clover silk, this parasite of many names is a relative of the morning glory. According to a late monograph, there are about 170 species of dodder belonging to the genus *Cuscuta*. Nearly forty that are injurious to agriculture grow in the United States. A number of the species are specialists, living as parasites on certain plants exclusively. In many places along the way, we saw the tangled masses of the threadlike stems splotching fields of clover with their color or overrunning weeds beside the road. Here among the green Nebraska dunes, the luxurious growth of the dodder reflected the greater-than-usual rainfall of the year. This annual plant thrives best during wet years.

Wherever it grows, its story is the same. From the sprouting seed, a slender thread pushes above the ground. It draws itself out to a length of from two to four inches. And as it elongates, it circles, groping blindly for a suitable host. If it fails to find one, it shrivels and dies. If it succeeds, it coils itself around the stem. Small wartlike suckers called *haustoria*, appear on the surface of the thread, penetrate the stem of the host and become fused with it just as a grafted twig becomes part of a tree. Through these the dodder absorbs food material from its host. Because it does not have to manufacture its own food, the dodder can dispense with roots as well as leaves. As soon as the parasite is well attached, the vine below the first coil dies. The plant has now lost all contact with the ground. Henceforth, by draining away fluid from its victims, it will thrive and grow, spread its tendrils from plant to plant, often end by forming tangled mats of threadlike yellow or orange-yellow stems many feet across.

During these midsummer days, the flowers of the dodder—minute and massed together, white or cream or pink, sometimes yellow—were preparing the way for the minute pods that would hold from two to four seeds apiece. As many as 3,000 seeds come from a single dodder vine. Brown, gray, yellow, reddish, they may be spherical or they may have flattened sides. Their exteriors are dull and slightly roughened. This latter fact has enabled commercial dealers to separate dodder from red-clover seeds by a treatment of iron powder, which adheres to the roughened surface of the dodder seeds but not to the smooth surface of the clover seeds. In a rapid mechanical sorting, all the seeds are passed over magnets and those of the parasite are eliminated.

These minute seeds, often overlooked, are distributed by irrigation water, in baled hay, among flax and alfalfa seed. They can pass uninjured through the digestive systems of animals. And they retain for years their ability to sprout. Many of the species that specialize in one particular kind of plant are synchronized with the host so the dodder seed sprouts only after the victim has appeared above ground and is well established. Even a severed fragment of a dodder vine retains for several days its power to coil and grow and develop new suckers.

An old-time superstition in the South had its origin in this ability of the parasite. Also the local name, love vine, apparently was similarly based. By swinging a fragment of love vine three times around the head and throwing it to the rear, the superstitious person sought to discover if a sweetheart was faithful. If, three days later, the dodder was found attached to some plant and growing—as was almost invariably the case—all was well. An even older superstition connected with this vine is recorded in the ancient herbals. A decoction produced by boiling fresh dodder vines was prescribed in past ages as a sovereign remedy for "melancholy, trembling of the heart and swooning."

GREEN DUNES

All through the dunes of Nebraska, on our last day there, we encountered that phenomenon of summer on the western plains, the passing showers of the one-cloud rains. Miles away we would see a cloud drifting toward us, trailing below it a smoke-gray veil of falling rain. It would pass overhead. Large drops would drum on the roof of the car. For half a mile or more we would ride on spattered or steaming pavements. Then sunshine would be all around us again and the narrow swath of the rain would sweep away across the countryside.

At times we would see as many as half a dozen widely spaced clouds come riding over the hills from the skyline, each with a pendant rain veil beneath it. Once, far to the south, where a heavy downpour descended in a gray column beneath a cloud, we glimpsed a thin line of lightning run for a split second, without a sound, down the center of the falling rain. Always the showers were of short duration. Always blue and sunny skies lay ahead and between the clouds. Later, in drier country to the south and west, we encountered other one-cloud rains, rains that never reached the earth. Resembling a school of rising jellyfish trailing filaments behind them, small clouds floated above attenuated veils of descending showers, rain that evaporated long before it reached the ground, sucked up by the thirsty air.

Clouds for a hundred miles around—angry, violent clouds, wind-torn and flushed in the last light of the sunset—marked the sky on the evening we came to the summit of Scotts Bluff. This sandstone eminence, beyond the dunes in the southwestern corner of Nebraska—famed in the annals of the Oregon Trail—rises a sheer 750 feet above the valley of the Platte. In the early hours of that evening, lightning glared all around the horizon. One-cloud storms were rushing through the sky, across the wide reaches of the plains. Beyond the Wyoming line, twenty miles away, a dark tower reared against the hearth-glow of the west, the black column of a falling deluge. Storm on storm advanced toward us. Yet before we left they almost all had spent themselves, and the serpentine of the Platte River, winding away off to the east, glistened faintly with the light of reflected stars.

Traveling downstream next day, past Chimney Rock and Jail Rock and Courthouse Rock, we rode on paved roads that paralleled the river up which the parade of the covered wagons, the Pony Express riders, the Argonauts, the Mormon pioneers, all the adventurous, restless, westward-streaming families, uprooted and going into a new, unknown land, had moved in a period of expansion. More than a century later, the ruts left by the heavy wheels of the Conestoga wagons are still visible in places along the Platte.

Down this river road, as well as along most of the summer roads we traveled, small birds, particularly house sparrows, would dash wildly across before us. At the last moment, as though on a suicidal impulse, many of them would plunge directly down into the path of the onrushing machine. Why did they descend into a zone of greater peril when they could have zoomed upward and let the speeding car shoot harmlessly past below? To do that, we decided, would have been a reversal of the age-old instincts of the bird—as much a reversal as for us to open our eyes wide instead of automatically blinking them shut when a dust cloud sweeps around us.

When it is hard pressed, the bird instinctively dives to increase its speed. It lets gravity add to its momentum. In darting across the road with a car bearing down on it, the small bird feels the need to cover the distance as soon as possible. It does what instinct tells it will shorten the time. It pitches downward. Again and again on our trip I saw small birds disappear in front of the car. I looked in the rear-view mirror expecting to see them killed on the highway. But they were not there. The burst of speed attained in their downward slant had carried them safely across. To fly upward and let the cars go by below would be safer still. But in the life of the bird, instincts are old and automobiles are new.

GREEN DUNES

Along this road beside the Platte, almost the exact day is known when the automobile first entered the lives of the resident birds. It was in the summer of 1903. Down this same highway we were traveling—then rutted and sandy, now hard-topped and smooth—a two-cylinder, chain-driven, twenty-horsepower Winton roadster had chugged eastward trailing a cloud of dust. At the wheel sat a young New England doctor, H. Nelson Jackson. He had made a wager of fifty dollars that, starting from San Francisco, he and a companion could reach New York under their own power. On this river road they were nearing their midway point.

The start had been made on May 23, 1903. Sleeping on the ground at night, crossing rivers on railroad bridges, following dry stream beds when roads were unavailable, pulling themselves out of mud as many as seventeen times in a day, they advanced eastward. Engineers stopped their trains to watch them go by, and cowboys rode their horses as far as 100 miles to see the horseless carriage. At Caldwell, Idaho, they took aboard a mongrel dog. Once Jackson's companion had to walk twenty-nine miles for gasoline. On a rough road in the Rockies, their front axle snapped in half. They put the two ends in a short length of iron pipe and drove twenty miles until they reached a blacksmith. When the bearings fell out of a front wheel, they replaced them with bearings from a mowing machine.

Sixty-three days after the start, they drove down Fifth Avenue, in New York City, riding on the same front tires and with the oil unchanged from start to finish. Their indomitable machine, with its wooden-spoked wheels and its fenders of plywood, the first automobile to cross the continent, is now one of the exhibits at the Smithsonian Institution, in Washington, D.C.

As we followed its pioneer trail eastward along the Platte toward Iowa that afternoon, we arrived at a point of special interest. Among all the land masses of the world, the continent of Asia is the highest. Its average elevation is 3,200 feet. The average for Europe is only 1,150 feet. The average for all the continents is 2,870 feet. Somewhere west of Kearney, between Lexington and Elm Creek, we came to one of the far-flung points that represent the average elevation for the North American continent. At that moment, our wheels rolled along the surface of the earth exactly 2,360 feet above the level of the sea.

TWENTY-TWO

GRASSHOPPER ROAD

A MILE of grasshoppers led us to the center of the United States.

At Grand Island we left the Platte and our route to Iowa, to drop south ten miles into Kansas. East of Smith Center, just above Lebanon, we came to a mile-long spur of hardtop. It extended across rolling land to a simple monument of field stones which, at that time, marked the exact geographical hub of the nation. Since then, the admission of Alaska as the forty-ninth state has carried this center point 439 miles north and west into upper South Dakota and the acceptance of Hawaii as the fiftieth state has moved it six miles west by southwest to its final location near Castle Rock, S.D., about twenty miles east of the point where the boundaries of Wyoming, Montana and South Dakota meet. But on this August day, the monument in the Kansas field marked the center of America. It also seemed, on that day, to mark the center of concentration of America's grasshoppers.

Nowhere else did we see locusts congregating so densely. All that mile of highway was covered with them. They paved the road with a second surface. They spread before us in a living carpet continually in motion. As we advanced slowly, thousands upon thousands rose, filling the air with their shining bodies. Like hail, they drummed on the metal roof of the car. When I stopped and stepped to the roadside, the vegetation appeared to dissolve into winged fragments as the multitude of insects swarmed upward with a dry roaring of wings around me. Once more we were in the presence of summer's overwhelming fecundity of insect life.

Before we reached the end of that mile of grasshoppers, the hood of our car was covered with the insects clinging in place by the adhesive pads of their feet. So firm is a grasshopper's grip on a smooth surface that it can maintain such a position on a car even at relatively high speeds. I remember once in North Dakota trying an experiment. I speeded up until I was traveling fifty miles an hour. A large grasshopper, facing straight ahead like an ornament near the front of the hood, still retained its hold. Chickens in the West have learned to come running when a car stops in a farmyard in the summertime. They recognize it as a source of clinging or dead grasshoppers.

Watching the locusts endlessly parading back and forth over the blistering pavement that afternoon, we noticed how, more and more as the day grew hotter, they tended to walk like crabs, with their bodies held high. Many seemed advancing on tiptoe. In these days when the fires of summer burn brightest, numerous insects thus reduce the temperature of their bodies by lifting them away from the heated ground, by letting more air circulate between them and the surface over which they are advancing.

Among the Indiana dunes I have watched field crickets venture out a little way onto the hot sand. Invariably their bodies appear lifted higher than usual. Certain leafhoppers in the desert regions of the Southwest possess abnormally long, thin legs that raise their bodies well above the ground when they have to walk about. Again, in the dry portions of the West, harvester ants can be seen lifting their abdomens higher and higher as the heat of the day increases.

Science has divided all living creatures into two groups, the *poikilotherms* and the *homeotherms*. The former, including the reptiles and the insects, are cold-blooded. They are unable to control their temperatures internally. Their bodies have roughly the same warmth as the medium that surrounds them. The *homeotherms*, comprising only the birds and the mammals, are warm-blooded. Their bodies maintain a relatively constant temperature the year around. It is interesting to note, in view of the bird's evolution from the reptile, that for the first nine days of its life a young passerine bird is unable to control its temperature. Its body heat may drop as much as 20 per cent when the parent bird leaves the nest.

Summer, for the cold-blooded, represents the Elysian days. Warmth brings life and animation. Their blood responds, literally, to every rise and fall of the mercury. Chill is synonymous with sluggishness; cold with immobility. The sun directly regulates the intensity with which they live.

In West Africa, the natives refer to the early morning sun as "the lizard's sun." For it is then, after the cool of the night, that all the small lizards are out basking in the warmth. Both lizards and grasshoppers have been observed regulating their body temperatures by changing their position in relation to the rays of the sun. Early in the morning, grasshoppers tend to place their bodies broadside to the sun, at right angles to its rays. In this position they gain the maximum effect of solar radiation. In this way, they sometimes raise the internal temperatures of their bodies as much as twenty degrees above that of the surrounding air. Like a man warming himself before a fireplace on a chilly day, the insects shift their position from time to time.

Later, when the sun has risen higher and the heat has mounted, the grasshoppers face directly toward or away from the sun, placing their bodies parallel to the rays. In this position they receive a minimum of solar radiation. Dark-colored species, naturally, warm up most rapidly. One entomologist found, by means of delicate electrical thermocouples, that during the same period of time the temperature of a brown grasshopper rose more than eight degrees F. above that of a bluff-colored one when the two basked side by side in the sun.

But even for cold-blooded creatures, stimulated by summer's warmth, there is a limit to the heat their bodies will endure. All along the way, as we roamed the mid-portion of the continent during the mid-portion of the season, we saw various creatures, cold-blooded and warm-blooded alike, adjusting themselves in diverse ways to the rigors of summer heat.

Fish descended into the cooler water of the deeper pools. Dogs lolled out their tongues. Rabbits panted. Unable to sweat, these animals cool their bodies by increasing the flow of air over their moist tongues and lips. Burrowing spiders were avoiding the heat by going to the bottoms of their silk-lined holes. So were ground squirrels and other tunneling rodents. In the depths of their holes, they spent the day surrounded by temperatures that rarely rose into the upper eighties no matter how hot the air grew outside. Gray squirrels, at times in summer heat, sprawl like miniature bearskin rugs on the cool, moist, open ground of shaded places. Once we passed fifteen tan piglets in a barnyard, all huddled together in the shade of a wagon. Where small digger wasps excavated tunnels in sand that was burning hot under the glare of the sun, they often flew about in the air to cool off after every few seconds of feverish burrowing.

By the time the mercury neared 100 degrees, we noticed all the perching birds let their wings hang down, holding them out from their sides. The plumage of a bird is less dense during the summer months. The creature is more lightly clad. In effect, birds wear lightweight suits in the summertime. A goldfinch, for example, has about 1,000 fewer feathers in summer than in winter. By wearing lighter colors during months when the sunshine is most intense, the collored lizard of Arizona reduces the effect of solar radiation. In winter it is dark-hued, absorbing more of the warmth of the sun; in summer its body becomes lighter in color, reflecting the sunshine.

Another creature of the desert, the kangaroo rat, has a unique water-cooling system that saves its life during peaks of abnormal heat. Unlike humans, who are fortunate in being able to perspire and thus cool their bodies, this rodent has no sweat glands. Instead it possesses a curious substitute. The highest internal temperature the kangaroo rat can endure is about 107 degrees F. As the animal approaches this lethal limit, a sudden and copious secretion of saliva occurs. It runs down its chin and the front of its body, wets the fur, and by its swift evaporation lowers the animal's temperature sufficiently to save its life. Research workers who have studied this emergency heat regulation have found that it may result in a loss of as much as 15 per cent of the creature's body water. This is perilously close to the 20 per cent which is considered the limit of desiccation that the animal can survive. In such circumstances, the desert dweller saves its life between two deadly hazards, approaching close to both—the hazard of heat that cannot be endured and the hazard of water loss beyond the point of recovery.

Because the relative surface area is greater in small animals than in large ones, the rate of evaporation increases as size decreases. This explains why the smaller creatures do not possess sweat glands. Their body fluids would be lost too rapidly. Loss of moisture is the great problem of summer heat. Insects have shown that they can stand markedly higher temperatures in moist surroundings than they can stand when the air is dry. The grasshoppers that swarmed over the monument road wore a thin outer coating of waxy material that reduced moisture loss through evaporation. Above about 111 degrees F., this insect wax begins to soften and melt, resulting, in very dry air, in a loss of body water. On the hottest days, grasshoppers climb vegetation to escape the heat being radiated from the ground. How swiftly this heat is dissipated is indicated by the fact that there may be a drop of twenty or even thirty degrees in the first half inch above an asphalt road, and a further reduction of several degrees for each additional inch. In times of great heat, the shaded sides of western fence-posts are sometimes fringed with densely packed masses of clinging grasshoppers.

It is in desert lands, dry and hot, that special adaptations for moderating the rigors of summer heat attain some of their most striking forms. One of the *Meloidae*, or oil beetles, of the Southwest, for example, has its elytra, or shards, fused together over its back, providing an air space, like an empty attic, above its abdomen. The insect has a swollen, enlarged appearance, suggesting half a miniature green orange. This globular shape presents the smallest surface to the sun, while the dead-air space beneath the shards provides a certain amount of insulation.

The strange Gila monster, the orange-and-black venomous lizard of the Sonoran desert of the Southwest, overcomes heat in another way. Birds nest early in this area. During the spring the Gila monster gorges itself on eggs. Its blunt tail swells into a larder of stored-up fat. Then, during the hottest time of summer, it lives on this accumulated food. Retreating to some hiding place, it falls into the torpor of estivation, that hot-weather counterpart of winter hibernation. This particular creature estivates in summer and hibernates in winter. Thus it spends much of the year in a state approaching suspended animation.

Various creatures, ranging from ground squirrels to snails, tide themselves over periods of intense, dry heat by sinking into a state of estivation. During this summer sleep, the functions of their bodies almost cease. Breathing is shallow, heartbeats slow down, digestive processes are almost suspended. So deep is its dormancy that the creature approaches closely to the border line of life and death. Thus it sleeps away the weeks of oppressive heat.

In Australia, R. J. Tillyard, the famous dragonfly expert,

discovered one nymph that estivates in the beds of evaporated pools and streams during the hottest, most arid weeks of the year. It can live for months in dry sand. In its dormant state, Tillyard reports in *The Biology of Dragonflies*, it becomes so desiccated it crackles when handled. Yet it recovers in a few minutes when placed in water.

A desert dweller that seems particularly able to endure great heat is the chuckwalla. This vegetarian lizard, sometimes attaining a length of a foot and a half, will rest for as long as twenty minutes unshaded from the burning midsummer sun. It can remain without any apparent discomfort on rocks that are too hot for human hands to touch. Among mammals, the antelope of the Great Plains gives the impression of being among the least affected by the blaze of the sun. It remains in the open when other creatures around it have sought the shade.

Shade is the lifesaver for a host of animals—including man. When we wear a hat or walk under a summer parasol or, for that matter, open the door and enter a house, we are making use of shade. So, in other ways, a myriad wild creatures turn to shade for relief from the midsummer sun. One noon, on a sun-drenched main street in a South Dakota town, we watched a house sparrow picking insects—grasshoppers and yellow butterflies mainly—from the fronts of parked automobiles. Each time it would flutter up, snatch a dead insect in its bill, then alight and hop rapidly across the sidewalk out of the torrid rays of the sun. There, in the shade of the stores, it would eat its prize at leisure before venturing out into the white glare once more.

Amid open rangeland, we saw cattle huddled in the sparse shade of cottonwoods and once, on the long, dry road down eastern Montana, we passed a forlorn sheep, with no tree to turn to, that was using the narrow shade of a fencepost. It stood with its body exposed to the burning sunshine but with its head carefully placed behind the upright post so the shadow strip ran across it. It seemed to be wearing the shade like a bonnet.

Mary Austin, in *The Land of Little Rain*, tells of a boundary fence that ran for a long way across rangeland in the Little Antelope country of the dry interior of southern California. "Along its fifteen miles of posts," she relates, "one could be sure of finding a bird or two in every strip of shadow; sometimes the sparrow and the hawk, with wings trailed and beaks parted, drooping in the white truce of noon."

Many times, as we rode across the open prairies, we saw birds perching on wire fences at the exact spots where the shadows of posts cut across them. Later, in northern Texas, a government naturalist told us of coming upon a line of fenceposts with a jack rabbit stretched out in the shade of each post. They all pointed outward with their backs against the upright wood, ready for an instant getaway if danger appeared. As the position of the sun slowly altered in the sky, swinging the post shadows over the ground, the animals kept shifting their places so they remained extended exactly within the narrow band of the outstretched shade.

The swiftest serpents in America are the racers and whipsnakes of the Southwestern deserts. They are day hunters. The slower snakes of the dry country come out mainly at night. For few reptiles can stand any prolonged exposure to the summer sun. The swift desert species streak from bush to bush, from shade to shade, their speed lessening the time they are exposed to the direct rays of the sunshine. In Death Valley, where the thermometer once reached the record height of 134 degrees F. in the shade, small harvester ants die in less than half a minute when they are out in the open during the hottest hours of the summer. Yet these insects continue to carry seedhusks to refuse piles outside the nest. But a scientist who timed these forays into the full sunshine found that the insects, in some way, gauged precisely the period they were without shade. The trips always ended before the danger point was reached.

Because, in the sandy stretches of desert country, the soaring heat of the day dissipates swiftly and is replaced by the cold of the night, many animals, particularly small cold-blooded creatures, avoid both extremes by becoming crepuscular. They emerge and feed in the twilight after the sun has sunk below the horizon. They are, in effect, using the shade of the planet to protect them from the heat of the summer sun.

In our own way, later that afternoon when we left the road of the grasshoppers behind and turned toward Nebraska again, we took advantage of the cooling shade. We were driving on a long, straight stretch of highway in the full blaze of the sun when we were overtaken by the moving shadow of a drifting cloud. I fed the engine a little more fuel and adjusted our speed to the pace of the shadow. For several miles we ran in this manner, with the sun veiled, bowling along at more than forty miles an hour, riding under the awning of a cloud. All the way there was the sense of a sudden lessening of the pressure of the midsummer heat. We wore the shade of the cloud as gratefully as the Montana sheep had worn the bonnet of the fencepost shadow.

TWENTY-THREE

THE CORN WIND

WE HAVE a midsummer but no midspring. We have a midwinter but no midautumn. Spring and fall are the seasons of the most obvious action and change. They are flowing this way and that continually. They have no long stagnant or slack-water times. But there comes a period in summer—and in winter, too—when day follows day with little variation. The wheel of the year appears slowing down. This, of course, is an illusion. For, as the poet Cowper observed long ago, all nature rides on the "unwearied wheel." But for the time being the illusion is strong and the progress of summer seems temporarily halted.

It was in such a time as this that we left Nebraska behind and entered Iowa. We crossed the wide Missouri going east, in the opposite direction from the westward-rolling pioneers. We crossed it on a bridge at Decatur, Nebraska, that was built in a manner new to our experience. At first it had been a bridge without a river. To reduce the cost by nearly half a million dollars, it had been constructed over dry land. Then the muddy flow of the Missouri was diverted under it. On the Iowa side, the structure ended in a long causeway across a lake that had been created from the pinched-off curve of the river's former channel. It was by way of this unusual bridge that we entered the land of corn.

Poets generally have sung of wildflowers and landscapes arranged by nature. It is the beauty of the unplanted and the untamed, of mountains and streams and forests and shores, that stirs us most deeply. But everywhere we went these summer days, Nellie and I appreciated, too, another kind of beauty in the out-of-doors, the beauty of cultivated fields and agricultural crops. By now we could close our eyes and conjure up their forms and colors—the red and green of blooming clover spangled with fluttering butterflies; rows of potatoes, all in bloom, extending for miles across the black loam of upper Minnesota; the North Dakota flax stretching away in a sea of blue; vineyards running in parallel lines up and down the Michigan hills; the great wheat fields, rippling, golden, cloud-dappled, restless as water in the wind. But none of the scenes that returned before our inward eye brought more delight than the remembrance of green corn, row on row, with banner leaves all flowing in the wind. This greater grass, the corn or maize, has a fluid, graceful, impressive beauty of its own.

All through Iowa—east to Storm Lake and south to Winterset and Oskaloosa—we were surrounded by corn, rarely out of sight of corn. Mile on mile, the rows went by, the great parade of corn, all drawn up in review. We saw the leaves shining as though waxed or varnished in morning sunshine. We saw them powdered with gray dust along the secondary roads. We saw the rows running up and over the hills, following the long straight lines laid down by the planters in spring.

In nature the shades of green are infinitely varied. But one of the most beautiful of all is the dark, rich green of a healthy cornfield. By observing tinges of purple or brown or white in the leaves, the expert can put his finger almost instantly on chemical deficiencies in the soil. From the color of the corn leaves alone he can determine if the land needs potash, nitrogen, phosphorus, calcium, magnesium or manganese. A strong purple tinting, for example, means a phosphorus deficiency. Brown along the edges or at the tips of the leaves indicates a lack of potassium. Here in Iowa the color of the corn reflected no chemical need that we could observe. This commonwealth contains a higher percentage of Grade A soil than any other state in America. Ninety-seven per cent of its area is under cultivation. It is the corn state pre-eminent.

During our days in this fertile commonwealth, whose name is derived from an Indian word meaning "The Beautiful Land," the corn was deeply green, the ears already formed. We saw the thousand-corridored fields in dawn mist and in brilliant sunshine and when the wind was blowing—the wind that adds its own fluid beauty to the cornfield. At such times the miles of corn, with leaves all flowing back in the breeze, looked as though it were marching across the land with banners flying. We seemed to be reviewing an immense army, extending away rank on rank across the state. Here was one vast parade ground of strength, the Iowa corn standing erect and tall.

Looking across the fields from an elevation, we sometimes noted how the green was topped with a golden-yellow wash. This was the color of the tassel fountains crowning the upright stalks. Viewed closely they suggested fingers outspread in the sun. Earlier in the season, innumerable, infinitely small bags of pollen dangling from these tassels had split open at the bottom to shower down the dust of fertilizing granules on the silk below. As many as 50,000,000 grains of pollen may descend, or be carried away by the wind, from a single tassel.

Like all the grasses, corn is self-fertilizing. The tassel represents the staminate flower, the silk the pistillate one. The threads of corn silk that extend beyond the husk at the tip of the ear are covered, at their outer ends, with fine hairs that catch the air-borne pollen. From a single particle the size of a dust grain, there develops a tube that travels down the center of the thread of corn silk to reach the ovary and produce the kernel. At times this fine tube has to extend itself downward for as much as eight or even ten inches. If no pollen reaches the outer tip of a thread, no kernel forms at its lower end. Each kernel is represented by a separate thread. When pollen is carried by the wind from corn of another variety and alights on a thread of silk, a kernel of a different kind results. It is in this manner that ears having kernels of several colors are sometimes produced.

In the seventeenth century, there was published in England an early account of the Virginia Colony that contained this description of the maize the white men found the Indians cultivating: "The graine is about the bignesse of our ordinary English peaze, and not much different in forme and shape, but of divers colours; some white, some red, some yellow and some blew. All of them yeelde a very sweete flowre; being used according to his kind, it maketh a very goode bread."

Nobody knows exactly where this grain originated or when it was first used as food. Such facts are part of the lore of ancient peoples. Maize is believed to have been cultivated first on the Mexican plateau. It is known that at one time the aborigines grew and used this grain as a staple food all the way from Peru to middle North America.

Its character apparently changed but little during all the pre-Columbian centuries. At Bat Cave, in New Mexico, archaeologists have found a pile of corncobs that they believe represents 4,000 years of feeding. The topmost cobs are just as stunted as those at the bottom. Forty centuries had brought about no improvement in the grain. It is only in very recent years, during the present generation, that the explosive advance of hybridization has changed this greatest of our native grasses into its most dramatic new and improved forms.

Almost every field we passed was marked along the roadside fences with metal signs: "DeKalb 3 x 1" or "DeKalb 3 x 2" or "DeKalb 3 x 3." We were surrounded by a world of hybrids. Twenty years ago, it is estimated, hardly 1 per cent of the corn planted was of hybrid varieties. Today more than 95 per cent of the nation's production is of this type. In two decades, as the result of scientific crossbreeding, the yield per acre has almost doubled. We saw the great ears swelling, already tilting down from the weight of their kernels. And over the hills and fields around us blew a summer wind, warm and gentle, a growing wind, the so-called corn wind of Iowa.

A spectacular feature of all these tall stalks that filled the landscape was the swiftness of their growth. The stems of most plants ascended through the multiplication and expansion of cells at their upper tips. They have a single point of growth. Corn, on the other hand, may have many points of growth. When it is young, it has growing centers not only at its upper tip but between each pair of joints on the lower stalk. It stretches out like the sliding legs of a tripod. It may develop from a kernel to a stalk twenty feet high between spring and September. In eight weeks, it may grow from a seed to a plant with 1,400 square inches of leaf surface. Its combined root system may total seven miles. We passed not far from one Iowa farm where, in the summer of 1946, the tallest corn in history grew. One stalk reached a height of thirty-one feet, three inches.

Like the man in the Grimm Brothers' fairy tale who could hear the grass grow, Iowa farmers are wont to maintain, half in earnest, that they can hear their corn grow on certain nights when both the temperature and the humidity are in the high eighties. On such a night, a stalk may add four inches, even six inches, to its height. In Iowa, toward the end of June when the mean night temperature rises above fifty-eight degrees F. and the mean day temperature rises above seventy degrees, the growth of the corn from day to day becomes perceptible. In *My Antonia*, Willa Cather gives a vivid picture of such a time when the corn shoots upward in the night.

It is this rapidity with which the corn grows and matures that enhances its value as an agricultural crop. John Fiske has pointed out that the swift maturing of the maize between the latest and the earliest frosts enabled the Pilgrim Fathers to survive during their first years in America. Here in the rich land of Iowa, the stalks that rose above our heads all had sprung up in a few weeks time, ascending in haste, elongating in spurts during the warm and humid nights of the early summer.

Many times, along the way, we drew up beside the great cornfields to examine more closely this plant that had played so important a part in the history of America. We looked at the slender height of each living tower of green. We looked at the hard, inelastic nodes or joints, closer together near the base, strengthening the stalk. Strong fibers extend from joint to joint like the cables of a suspension bridge. The farther-apart spacing of the joints near the top permits greater bending and flexibility. This combination of hard nodes and flexible sections between them enables the corn to survive against the repeated attacks of its age-old enemy, the gusty wind of summer storms.

The leaves, too, are designed with long, almost parallel, strengthening veins. Down their length, bisecting the rich green with a central line of lighter color, runs the strong backbone of the midrib. Strengthened thus, the corn leaf is free and fluid in its movements. It is fitted to wave and stream and whip in the wind without breaking or dragging over the stalk that supports it. Each leaf is anchored to its stalk at one of the hard bracing joints, thus giving it firm support. The all-important ears, the seed of the corn, are also always formed at these strong points on the stem. They are surrounded and protected by a husk of modified leaves. The cob is essentially a shortened stalk that juts out supporting the close-packed kernels of the ear.

At the base of every leaf there is a rain guard that prevents water from seeping down between the stalk and the leaf where it would aid in the growth of fungi. And near the base, folds on either side enable the leaf to swing sidewise without damage. In times of drought the corn leaf rolls itself into a long tube, lessening the surface it presents to the air and the wind and, in this way, reducing the transpiration of water. With the coming of rain, the leaf unrolls and normal evaporation of moisture into the air is resumed.

While the stalk and leaves of the corn plant are designed to meet with resiliency the force of the wind, the roots at the base are fitted to give the high, slender stem ample anchorage. The main roots plumb the earth to a depth of as much as six feet. But in addition there is an outspreading web of bracing roots, extending away close to the surface of the ground and functioning like stay ropes about a flagpole. They keep the stalk from being knocked flat in high winds. Even when the wind succeeds, when a stalk lies prostrate in the wake of a gale, the resources of the corn are not exhausted. On the lower side of each node or joint, a growing wedge begins to form. As these wedges enlarge, they force the stalk in a rising curve upward toward the sun. The plant never becomes vertical but, in this manner, it is able to lift most of its length from the ground and assume a growing position again.

Of all the plants of the earth, the grasses give off the most moisture into the atmosphere. At the height of its growing season, in late June, an acre of lush meadow may, in a single day, transpire into the atmosphere as much as six and a half tons of water. One stalk of corn, on a summer day, will raise and disperse into the air as much as a gallon of water. During its growing season, an acre of corn may evaporate into the air above it 300,000 gallons. It has been calculated that if all the water expelled by the leaves of the corn were collected into a lake, it would stand five feet deep across the length and breadth of the cornfield.

"Corn," an old saying has it, "takes a lot out of the soil and never puts anything back in." It is a thirsty drinker and a lavish feeder. Here, where corn ran beside us wherever we went in those early-August days, it was obtaining the nutrients and moisture it needed from within the rich Iowa soil. That soil, laid down by river sediment through the ages, is the great

THUNDERHEAD lifting its tumbling vapor high into the summer sky above a field where the wheat is ripening.

RIM OF THE WORLD, the flat land of western Kansas. Below, a buffalo rubbing rock on the Dakota prairie.

bank of this Midwestern state from which withdrawals have been continually made.

The air at Storm Lake, when we stayed there overnight, was filled with a strange chemical smell. It came from the big processing plant where the DeKalb company treats the seed kernels of its hybrid corn. Out in the country, several times a day, we passed experimental fields in which new types of hybrids were being developed through scientifically controlled pollination. The search still continues for new improvements and higher refinements in *Zea mays*, that far-away parent plant cultivated by the aborigines.

In these summer days, workmen were already cleaning up the grounds and buildings where country fairs would be held. Summer auctions of livestock and household goods were in full swing and we could hear the stentorian voices of the auctioneers as we rode by. For more than 250 miles, we angled south and east through corn country. Nighthawks and swifts were calling under an overcast sky the morning we came to Oskaloosa. Its name, derived from that of the wife of the great Indian chief, Osceola, means "The Last of the Beautiful." Once briefly, during summer days many years before, I had visited this Iowa town. Here I had first heard Hawaiian music, and here as a boy I had read my first book by Dickens, *Oliver Twist*. Under the swifts and nighthawks, on this day, I drove about as I sought dimly remembered former scenes. That noon we ate at a restaurant that served a deliciously flavored, wonderfully cooling dessert we had never tasted before—watermelon sherbet.

The heat mounted during the next two days as we descended south into Missouri and curved west into Kansas. One of the last things we saw beside the cornfields of Iowa was a goldfinch, chosen officially as the state bird. And one of the first things we saw after we entered Missouri was a bluebird, the choice in that commonwealth. It was perching on a telephone wire above great purple plumes of the blazing star or gay-feather that rose densely for an eighth of a mile beside the road.

All through our summer journey, as we advanced from state to state, we saw the succession of beautiful birds that have been chosen, officially or unofficially, by the various commonwealths. They ranged from the purple finch of New Hampshire through the hermit thrush of Vermont, the cardinal of Ohio, the robin of Michigan, the western meadow lark of Montana, the scissor-tailed flycatcher of Oklahoma to the lark bunting of Colorado.

Only once before, in late-summer days more than thirty years previously, had we ridden this way—westward through Missouri into Kansas and down the long southward slide to Wichita. It was the year Nellie graduated from college, the year we were married. I was returning for my second and Nellie was starting her first year teaching at Friends University. We went west in a Model T Ford and camped along the way. We carried our khaki tent rolled up in a luggage rack formed by a lazytongs bracket that clamped to a running board. In Illinois we plowed through gumbo mud almost hub deep. In Missouri we crossed one deep creek ravine on planks that sagged with the weight of the car. And for the last hundred miles or more, as we neared our destination, we advanced over the long swells of an endless sea of prairie sunflowers.

The automobiles we own, like larger calendars, divide our lives into segments or compartments. We tend to date events by our cars. "That was when we had the Model T," we say, or "That was when we had the Model A," or "That was when we had the Chevrolet" or "That was when we went in the Buick." Of the cars with which our lives have been intimately associated, that black Model T stands out most vividly in memory. It was not only our first car; it was an exacting and temperamental personality. We ran all day with our headlights on to keep the generator from burning out. We rode with the hood off most of the time to keep the engine from overheating. Going downhill we used the reverse pedal as a drag to prevent the brakes from burning out. We cranked the car with care lest the handle kick back and break a wrist. And when the motor caught, we raced back to the steering wheel to jerk down the accelerator lever before the engine stalled. To determine how much gasoline we had left, we took off the front seat, unscrewed the cap of the fuel tank and thrust in a stick. When it rained, we got out the side curtains and snapped them into place, usually waiting until the last minute and being drenched in the process.

That first westward trip was made in the mid-twenties. It was made over dirt roads, in clouds of dust, running on good stretches at thirty miles an hour. Now with automatic gearshift and power steering, riding on a superhighway where signs announced a speed limit of eighty miles an hour, we sped toward our destination. As we advanced we talked of that earlier trip in another summer. At that time all the events of the subsequent years lay beyond predicting. Now we knew chapters in our lives then unwritten. As we rode on, memory revived minute and forgotten details of that former journey we had made a third of a century before. And when at last we came within sight of Wichita, we felt like pioneers returning.

TWENTY-FOUR

TEN THOUSAND STEPS

THE five boles of the willow tree stood apart like the ribs of a partially opened fan. Three of the five trunks leaned outward from the river bank, the slow swirls and eddies of a thin brown current sliding away downstream beneath them. At the end of this August day, while Nellie rested from the heat, I ate a sandwich and drank a pint of milk on the sandy bank beside the willow. I watched the flotsam of the stream—yellow leaves, newly fallen; old leaves, decayed and almost black—go drifting by.

More than thirty years before, just out of college, a stranger in a strange land, teaching at Friends University students almost as old as I was, I used to spend the precious outdoor days of my week-ends tramping along this very bank of the Arkansas River south of Wichita. On a little island, not far from where the willow stood, I had cooked a small steak over a campfire for my Thanksgiving dinner and, sitting on a stranded log before the fire that afternoon, had finished Joseph Conrad's *The Shadow Line*. That first year the river bed had been so dry I could follow it like a broad highway far into the country. Usually on these expeditions a sandwich rode in one coat pocket while some small book, such as Amiel's *Journal* or *The Golden Treasury* or Hugo's *The Laughing Man* or Yeats' *The Land of Heart's Desire*, occupied the other.

On this day, under the willow, the book I pulled from my pocket was only five and a half inches high and three and a quarter inches wide. It was a copy of James Thomson's *The Seasons* printed in Boston in 1823. Bound in brown leather and scuffed from much usage, it had ridden with us on our spring journey. Now it was our companion in summer. More than two centuries after this English classic of nature poetry first had been set down, it was accompanying us through all the seasons of the American year.

When I closed the book, I looked again at the slow eddies turning on the brown current, new swirls forming where other swirls had been. Where were the islands I had known three decades before? Where were their particles now? In this ever-changing stream, new islands are a measure of the years. I gazed down at the flow of the water, the sediment and sand grains moving downstream with the current. Since last I stood here, the days of three decades had been eroded away like island sand. Time is the river. We are the islands. Time washes around us and flows away and with it flow fragments of our lives. So, little by little, each island shrinks. In the waters of the Arkansas, eroded particles may contribute, far downstream, to the growth of another island. They are deposited elsewhere along the flow of the river. But where, who can say, down the long stream of time, are our eroded days deposited?

I followed the bank slowly southward. The river flowed beside me as all streams move—faster at the center, swifter at the top, the water along the banks and close to the bottom slowed by the greater drag of friction. A twinkling of minnows glinted in the muddy shallows. Not far apart along the bank, a frog shot into the water, hidden instantly in a cloud of erupting sediment, and a bright-eyed little lined lizard, with a wash of yellow along its side, streaked up the sandy slope. For it, there was no frog's muddy hiding place. For it, safety lay in its spurt of speed.

South of the city and north of the city, the next day, beyond the farthest reach of the subdivisions crowding the banks, I wandered along this remembered river. In all I walked, perhaps, five miles or more—ten thousand steps at our average rate of two thousand steps to the mile. All along the way, here where earlier days had been spent intensely, moments of the past sprang suddenly into life from the subsoil of my memory.

Wichita, as I recalled it, had been a town of 60,000. Now it was the largest city in Kansas. Its population was nearing the quarter-million mark. New bridges spanned the river. There were houses, houses everywhere. All seemed changed. Yet when is a city without change? A city is no more static than a stream. Never on two successive days do we find the components identical. A birth here, a death there; a train discharging passengers, a train taking on passengers—thus the population of a metropolis alters slightly almost every hour of the night and day.

I searched in vain along the river banks for some of the places I remembered well. Where had I listened to the great chorus of frogs in the soft air of one spring evening long ago? And somewhere here, I could not tell where in a world so changed, I had stood on a low hillock of sand one smoky autumn dusk and looked away across the darkening landscape to a mysterious house that stood alone, crouching like a gray cat in the midst of a wasteland, returning my stare with the two yellow eyes of its lighted windows.

Although Kansas is a comparatively dry state, it contains several thousand streams large enough to be named. At the head of the list, the largest to flow across its land is the Arkansas, the river beside which I walked. Born among snowfields above Leadville, Colorado, high in the Rockies near Fremont Divide, it drops 5,000 feet in its first 125 miles. Along its upper reaches, in the gold rush of 1859, prospectors struck some of their richest deposits of precious ore. Spectacular Royal Gorge, a product of its torrent, now walls in the millrace of its waters with sheer cliffs that rise as much as half a mile straight up on either side. Escaping from the mountains, this greatest of the Rockies' east-slope streams flows across the plains of Colorado, winds for 402 of its 1,450 miles through southern Kansas, cuts across the northeastern corner of Oklahoma and traverses Arkansas before it joins the Mississippi. Again and again, in days that followed, we crossed the river farther up its course, in western Kansas, on the high plains of eastern Colorado, at the Royal Gorge, near Leadville and the beginning of its flow.

In many ways it is a remarkable stream. It is the only river I know that, midway in its course, changes the pronunciation of its name. From its source to the Oklahoma line it is known as the "Ar-kansas" River. From the Oklahoma line to its juncture with the Mississippi it is the "Arkans*aw*." That pronunciation has been required by law in the state of Arkansas ever since an act of the legislature in 1881. Down the length of its course, the river drains about 185,000 square miles and by the time it reaches the Mississippi it has descended more than 10,000 feet below its starting point close to the Great Divide.

Reduced now by heat and summer drought, the river beside me was a leisurely and peaceful stream. It wound about in its own wide, sandy bed, a far cry from rocky, rushing Sunday River in Maine. Yet here, too, was the flood wrack of the spring torrents, bleached and battered barkless trees strewn along the banks. I sat on one stranded cottonwood, half buried in the sand, and watched particles of rock and sediment being inched along by the current below me. Where had they all come from, these grains of quartz and granite, these minute fragments of rock, these specks of humus that rode with the water or were pushed along the bottom? Some had been part of the high, dry plains to the west at a time when neither Indian nor bison had ever seen a white man. Others represented grist from the mills of the glaciers, infinitesimally small subtractions from the bulk of the Rockies.

More than 400 years before, in the summer of 1541, Francisco Vasquez de Coronado had led his gold-hungry band around the great bend of the Arkansas and followed the river downstream, it is believed, as far as the present site of Wichita. Along this very stretch of the river where I clambered over tangled driftwood or rested in the shade of the willows, his men may well have built their campfires of similar flood waste and found temporary shelter from the sun under just such willows.

For, like the cottonwoods, willows are an important feature of the rivers of the plains. They survive in heat and cold. They prosper under adverse conditions. "Ah, willow! willow! Would that I always possessed thy good spirits," Thoreau wrote in the last volume of his *Journal* when his own life was nearing its end. All the way from the Rio Grande to the Arctic Circle, the seventy or so species found in North America survive amid an infinite variety of conditions. On the Great Plains willows are employed as windbreaks, and along the levees of lowland rivers they are planted to prevent flood damage. Growing rapidly, willows in some favorable locations have become large trees in five year's time.

Along the streams of the eastern states, black willows sometimes attain a height of 120 feet. Yet above timberline in the Colorado Rockies, at an elevation of nearly 12,000 feet, we found stunted willows, covered with the white fluff of their catkins, that were only four inches high. Among the Yukon Indians the inner bark of willows is eaten as a vegetable food, while above the Arctic Circle Eskimos store leaves and shoots in oil, adding them to their winter diet to prevent scurvy. It was from the bark of the willow that science extracted the original ingredient that led to that medicinal boon, aspirin.

Somewhere beside the river, as I followed a once-familiar way now become alien and strange, I spied an old newspaper discarded among the sandburs and sunflowers. Symbolic of the profound changes of the years, of the new age in Wichita, it was *The Wall Street Journal*. Beyond, I had taken no more than a dozen steps when, in extreme contrast, I came upon a battle in the dust so primitive it might have been occurring millenniums before Coronado sailed from Spain.

Across the sandy path, red harvester ants were dragging a greenish caterpillar. They walked backward, tugging at their burden, hauling it laboriously with starts and stops toward the entrance of their nest. Just as I arrived, a black *Calosoma* ground beetle, a caterpillar hunter with a body an inch long, discovered the ants and their prize. It rushed forward on slender, nimble legs. Grasping the larva in its powerful mandibles, it attempted to snatch it away from the half dozen clinging ants. They held on. Other ants swarmed to the attack. They climbed the legs of the enemy and sank home their jaws, seeming to concentrate on the joints. The *Calosoma* hastily dropped the caterpillar, shook itself free and retreated.

For a minute or more it prowled about out of reach of the ants. Then it darted in a second time. And a second time it was repulsed in the same way. Three times I saw it repeat the attack. Each time it met with such stout resistance it was forced to give way. After the third defeat, it wandered off up the path seeking other and easier prey, its meandering course recorded by the delicate ribbon of tracks left by its feet in the dust.

Nearly as alien as the paths beside the river were many of the streets that Nellie and I thought we remembered well. So many things had changed, so much had happened to us in the intervening years that, as we drove about in the twilight, we seemed revisiting scenes in some reincarnation. The great red brick bulk and the high tower of Friends University remained an unchanged landmark. We remembered it with affection. There we had both taught during my second and last year in Wichita. Before we realized it we were riding on Vine Street, now Vine Avenue and a main artery of the city, and before we recognized it we were passing the white frame house where we had lived in three upper rooms that first year of our married life. Below us the owner, a kindly maiden lady, Myra Lunt, had dwelt with a cat named Thomas-Elizabeth. Originally it had been named Thomas. The Elizabeth had been added somewhat hastily when the kittens came.

To the south of this house, across interurban trolley tracks now gone, there used to stretch weed fields and vacant lots. There, long after midnight, surrounded by black, silent shadows, I once conversed in low tones with a ghostly figure whose face I never saw, a man who, although we stood within touching distance, seemed separated by an enormous gulf. Remembered through the years, it was a strange, almost eerie experience.

I had returned, as coach of the college debating team, on a late train from Newton, Kansas. About a block from home, I was walking down still, deeply shaded Vine Street when a slender man stepped from the shadows into the light of a corner lamp. His hatbrim was pulled down, his raincoat collar turned up as he walked across the street. The light glinted on a revolver held in his hand. One pace behind me, his gun prodding my back, he marched me past the darkened and sleeping houses, past the dog that barked and barked and awakened nobody, across the interurban tracks and on out into the deserted expanse of the weed lots. There, with the muzzle of the revolver pressed between my shoulder blades, he carefully went through my pockets, extracting my wallet, my watch, my fountain pen.

And what was I thinking at the time? I have tried to remember. There were many things. But mainly I was concerned about the chill of the night, the fact I had on only a light suit, that I had been shivering before I met the highwayman and now my teeth were chattering. That irritated me greatly. I knew inside I honestly was not *that* scared. The second thought that grew as the first shock wore off was that this would probably be the only time in my life that I could talk to a highwayman, it was my only chance for a glimpse into his world, so different from my own. That immense curiosity about what life is like for the squirrel and the robin and the earthworm, what existence means to them, extended to the shadowy, desperate man behind me. His mind represented a foreign shore, so near and yet so far away.

By holding me up, I pointed out, he was taking about twenty dollars. If he got caught he might be sent to the penitentiary for twenty years. Why take such chances? His answer—which may have been something of a stock alibi—was that he had been sent to prison for a crime he hadn't committed; when he came out he had a record; the police hounded him so he couldn't hold a job anywhere. He said I must have read about him in the papers. He was the Midnight Bandit that had been in the headlines all over Kansas and northern Oklahoma. He seemed annoyed that I had not heard of him. He was a celebrity. The papers were talking about *him*. The exploits they reported were *his* exploits. Yet only at such moments as this, in shadows, surreptitiously, with the listener covered by the unwinking eye of his revolver, could he reveal the truth. His success was secret, his career anonymous. To victims alone could he relate his exploits. He was really living only at night; he passed through his days incognito.

I have often thought of him since, of the urgency with which he exhibited that human trait shared by great and small, the desire to have someone else know of his achievements. So Napoleon Bonaparte, as Emperor of France, wondered if his old mathematics teacher, who thought so little of his ability as a schoolboy, had heard of his renown. Most of all this human foible concerns the people who knew us long ago. I remember interviewing for a magazine a man who had been born in a small hamlet in northern Georgia and who later had attracted wide attention as a handler of venomous snakes at Ross Allen's Reptile Institute, at Silver Springs, Florida.

"You know," he told me as I was leaving, "I really don't care much about having my story in a national magazine. What I'd like is for somebody to write a piece about me for a little weekly newspaper up in Georgia. I'd like the folks back there to know what I've amounted to!"

During our days in the vicinity of Wichita the city steamed in an August heat wave. Each afternoon the thermometer climbed to around 100 degrees. Thomson describes such a time as this in *The Seasons:* "Distressful Nature pants. The very streams look languid from afar." Then, during the last hours of our stay, in one swift and violent overturn, the heat was lessened and the drought was gone.

Late that afternoon, a few miles west of the outskirts, we turned north in search of the river. Our road ran on and on straight before us, first hard-surfaced, then double tracks in sand and dust. It led us into wild country and into wild and memorable weather. There was, when we started, a sense of storm in the heavy, breathless air. To the west, clouds rolled upward. They darkened to purple. Lightning split long cracks down this swollen sky. Suddenly, chill air was around us and dust was sweeping in cloud after cloud down the road. A cold front was breaking the heat wave with a storm.

We had gone ten miles or so before we came to the river. Here it was no tamed and channeled stream of the city. Here it was wide and wandering. An eighth of a mile apart, two levee-like ridges of sand had been thrown up, one on either side, to confine the flood waters of spring. But otherwise, here at last was the lonely river, the Arkansas I remembered.

Around me, as I stood on the top of the nearer levee, the wind accelerated swiftly. It roared among the branches of the half-dead cottonwoods. Leaves scudded past. Wherever my skin was exposed, it stung with the lash of gust-driven sand. I stood there with the stirring march of the storm wind in my ears. When I struggled down the side of the levee again a dusty-colored toad hopped before me. In the sand near the car I came upon another toad. Close by, where the embankment had been cut vertically by a passing road scraper, I caught sight of the doorways of several exposed miniature caverns. In one, a third toad was looking out. In another I saw a dim shape, with dark and yellow mottlings, crowding back into its shallow room. I leaned closer. A loud hiss greeted me. Within its modest cave in the sand a land tortoise was awakening. Toad and tortoise—these and how many other humble creatures of the dust were being aroused into new activity by the approach of the summer rain!

The downpour came with a rush. In a swift curtaining of the landscape the sheets of water fell. For a quarter of an hour we were in the midst of a full-fledged cloudburst. Then the great deluge slackened. A half an hour later the wind had lessened. But all through the evening and into the night the rain fell in a steady shower.

Returning through a wet and dripping world, we were surrounded everywhere by a resurgence of life. Everything seemed to feel exhilarated by the storm—and so did we. This was the summer rain. This was the breaking of days of drought. This was the conquest of the heat wave. Plumage and skin and the chitin shell of the insect all were washed clean once more. The dust was laid. Life responded, each form in its own way. We stopped once beside a clump of willow trees where a whole family of dickcissels were darting and calling among the branches in a kind of ecstasy. They and other birds along the way seemed stimulated by the storm. But it was when we crossed a mile or more of lowland meadows that we found ourselves amid the most dramatic manifestation of all.

The road was streaming with water. And everywhere along its length hopped an incalculable multitude of little toads. It was easy to understand the old mistaken belief that toads fell from the sky with rain. All down the sandy road we had seen others in smaller numbers. But here on this lowland hardtop the toadlets were in such throngs that at times the wet road seemed alive and moving. Once we saw four—and again five—leaping shoulder to shoulder, jump for jump, as though they were running races. Dusty toads were dusty no more. Days of hiding were, for the time being, over. This was their Fourth of July, their Liberation Day. They seemed in the midst of a great celebration.

So we came back from the river in the rain. We returned elated, our spirits soaring as though we, ourselves, had caught some of the infectious joy of this multitude of toadlets suddenly set free.

TWENTY-FIVE

NIGHT OF THE FALLING STARS

THE dust of the lonely road lay pale silvery-yellow in the moonlight. Its narrowing ribbon extended away before us across a land level and dark and outspread beneath the immense glitter of the prairie sky. It was two o'clock in the morning. We were a hundred miles west of Wichita, a hundred miles from the joyous toadlets splashing in the summer rain. Here no drop had fallen for weeks. Here fields were parched; farmhouses far-scattered across an arid land. No dogs barked in the night. Standing there amid the silence of the darkened prairie, we were surrounded by a scene that was nine-tenths sky. We felt, in this hill-less, treeless region, as though we were in direct contact with the heavens, at a meeting place, on the sky-shore. The everyday life of the world had receded far away. We were alone among the stars.

And this was a night of nights to see the full sweep of the sky unimpeded by house or hill or tree. For we were now in the time of falling stars, the time of the summer Perseids, the greatest meteor show of the year. During the second week of August, each year, the orbit of the earth carries it through a celestial cloud of fragmentary matter streaming across the void at planetary speeds. Fragment after fragment, turned incandescent by friction in the earth's atmosphere, draws in a swift, bright stroke a line of fire on the firmament. For hours now, in spite of the light of a moon almost full, we had been seeing these dark stars from outer space burst into transitory brilliance and imprint for the space of seconds their flaming paths across the dome above us. Where had they been only yesterday? Farther away than the moon. Where were they two months ago? Perhaps as remote as the sun.

Most of the shooting stars we saw that night streamed from the northeast, from the direction of the constellation Perseus; hence the name "Perseid showers" for these celestial displays of August. Each rushing meteor enters the earth's atmosphere at a speed of about twenty miles a second. It becomes luminous at a height of about sixty miles and, as a rule, vanishes, burned out, at a height of forty miles. The most brilliant of all meteors, bright enough to cast a shadow, are called fireballs. It is the incandescent object moving through the atmosphere that is the meteor. The stony or metallic object that occasionally survives and reaches the earth is a meteorite.

When we started out, about 10:30, the celestial show had already begun. But, as always, it was after midnight before it reached its height. Then four or five times as many meteors are visible in the heavens. After midnight we occupy the forward side of the turning globe as it sweeps through space. We encounter the meteors head-on. Their speed and brilliance increase. In the evening hours we are at the rear of the globe. The shooting stars we see are only those that overtake us.

Standing there in the dusty road, we watched these fragments from outer space consume themselves high in the atmosphere. Their sudden trajectories lengthened and disappeared above us. We watched them drawing fiery lines across Cassiopeia and the Big Dipper. At their peak, on a clear, moonless night, the meteors of the Perseids crisscross in the sky. On this night the light of the moon no doubt competed, erasing the fainter, more distant trails. We would see three or four streaks across the sky, one coming after the other. Then there would be a lull; then a long single line, sometimes soaring overhead, sometimes cutting to the right, sometimes

PRAIRIE DOGS in a friendly meeting below the great monolith of Devils Tower in northeastern Wyoming.

VISITING prairie dogs, above. Below left, a prairie dog eating. Right, pausing alertly at burrow entrance.

AT EASE, this prairie dog sits upright beside its burrow. Following page, emerging into the sunshine.

NIGHT OF FALLING STARS 253
descending in a shallow arc to the left.

In any twenty-four hours, it has been calculated, at least 10,000,000 fragments from outer space enter the earth's atmosphere. The number has sometimes been placed as high as 4,000,000,000. Yet one astronomer computes that their average total weight would be hardly more than a ton. Most are so minute they disappear almost instantly. However, a bit of matter no larger than a pinhead will produce a faint, short-lived streak in the sky. These are the smallest objects in the solar system ever seen by the human eye.

As the hours of that night wore on, we wandered about over the level land, changing our viewpoint, standing in the dust of the side roads while the succession of quick, flaming trails shone in self-consuming brilliance in the sky above us. Several times we passed a low, darkened farmhouse, its cyclone cellar hidden in the shadows, its two weathered windmills motionless in the moonlight. Once we looked back as we turned slowly down a crossroad and saw what we thought was the most brilliant meteor of the night trace its path half across the sky, sinking gradually and fading away, from our point of view, directly over the old farmhouse. Just so, perhaps, on some long-ago night, out of just such a glitter of stars, had come to the dark acres of the surrounding land a shower of wealth from outer space.

For these acres in Kiowa County, Kansas, hold a special place in the annals of meteorite history. About one-third of all meteorites found in North America, and almost one-sixth of those recorded throughout the entire world, have come from Kansas. The plowed fields of the western half of the state have been especially productive. But none contributed more stones from the sky than this land that bordered our lonely road. It comprised a tract that has been famous for more than a half a century as the Kansas Meteorite Farm.

In 1885, a young couple from southern Iowa, Frank and Eliza Kimberly, moved here, six or seven miles from the present town of Haviland. The earliest pioneers in the region had noted that the only rocks they found were heavy and black and strewn about in a relatively small area. Homesteaders used them to weight down haystacks and fences and the covers of rain barrels. They were also employed to anchor in place the roofs of the dugouts and protect them from the violence of the wind. But nobody suspected the dramatic history of the stones until Mrs. Kimberly picked one up not long after she arrived in Kiowa County.

As a small girl in Iowa, she had been taken by a teacher to see a large meteorite that was being taken through town on its way to an Eastern museum. She now recognized the object in her hand as similar to it. Day after day she hauled other heavy, black stones in from the fields. As her pile grew, she began writing to scientists trying to interest them in her find. But her farm was remote and they were skeptical. Years went by without any response. Her pile of black stones became a standing joke throughout the region. Nevertheless, year by year, she added to it.

Then, in 1890, Dr. Cragen, the head of the geology department at Washburn College, in Topeka, drove out to the pioneer home where the Kimberlys lived and was dumbfounded to discover something like a ton of piled-up meteorites. He bought some, museums bought others, and from the stone pile that was no longer the joke of the region Mrs. Kimberly realized enough to pay off the mortgage on the farm. Not only that, but she was able to buy the adjoining farm as well. When, before she died, she had become the richest woman in Kiowa County, there were no hard feelings over the money her meteorites had brought her. She was considered by her neighbors "a mighty smart woman," an opinion that was not lessened when it was learned that she had paid a lawyer one dollar for a meteorite she later sold for $500.

That afternoon, while Nellie was getting ready for our night with the stars, I had driven west from Pratt, past the

NIGHT OF FALLING STARS

dry little town of Haviland and off on dirt roads that intersect at mile intervals and cut the plains into innumerable squares of equal size. At the second crossroad I had swung west and come to the meteorite farm. Here Kimberlys still live—Oren Kimberly, a middle-aged, stocky farmer who wore a long-peaked cap indoors and out and used slow, humorous speech, and his mother, Cora Kimberly, Eliza's daughter-in-law, a slender wisp of a woman in her early eighties who had lived in this same house for more than sixty years.

Looking down from the wall, as they recalled earlier events that had brought fame to their Kansas acres, was an enlarged photograph of Eliza Kimberly. Hers was an interesting face: capable, intelligent. She was, I surmised, the kind of person who would appreciate the ironic humor of the official records that list the great discovery of her meteorites as occurring in June, 1890—that is, when a scientist finally arrived and discovered what she had discovered and the meteorites she had collected years before.

As Nellie and I watched the falling stars that night, we wondered if any we saw were reaching the earth. It is, of course, only the largest masses that survive the fall. During its long descent, a meteorite may lose as much as half its material through heat and abrasion. Stony meteorites, by far the most common kind, are almost always characterized by a black fused crust, the product of intense heat in the sky. So far as is known, the largest ever to reach the earth intact lies buried amid limestone near the town of Grootfontein, in Southwest Africa.

Somewhere in the darkened fields that spread around us, one of the biggest of all the Kansas meteorites had been discovered in the early days by the Kimberly's hired man, Jack Sanders. One evening he came in from a day watching the cattle and told Mrs. Kimberly:

"'Liza, I saw one of those black stones you've been saving, today."

They all trooped out to look at it. But Sanders couldn't find it.

"If you ever see it again," Mrs. Kimberly told him, "you sit right down on it. Let the cattle go. Stay right there until we come and find you."

A few weeks later the Kimberlys discovered their cattle scattered all over the fields and no Jack in sight. They hunted him up and found him sitting on the ground beside a small point of black rock projecting above the surface and half hidden in the grass. Digging down at this spot they unearthed a stony meteorite that weighed 700 pounds.

At first her husband viewed her growing stone pile with tolerant humor. Later, when its value became apparent, he hastily hunted over the fields for a large meteorite he had once discarded. His wife had found it in a remote part of the farm and had put it, with several others, in an empty wagon for him to bring to the house. He decided she had enough stones and heaved it into the bushes somewhere along the way. Hunt as he would he could never find it again. Later in our trip when, in the high Rockies, Nellie became enamored of rocks and loaded the car with specimens until the springs bent and the trunk sank low, there were times when I extended a surreptitious hand to lighten the load, but, just in time, I always remembered Mr. Kimberly and his thrown-away prize.

Although Eliza Kimberly never owned a telescope, she was long interested in the study of the stars and the constellations. Her eyes, I was told, were unusually far-sighted. On this August night, above our heads glittered the same stars, the same westward-wheeling constellations, that those far-sighted eyes had seen. From this identical point on the spinning globe where she had watched the procession of the nocturnal sky, we were viewing it on this night of the Perseids.

For more than six decades the fields around us had continued to yield new finds. Most of them are now scattered among American museums. A few have gone to European institutions or, as Oren put it, "way over in those foreign countries." Not a few of the Kimberly meteorites have been discovered accidentally. In 1925, Oren located one that weighed 465 pounds when his plowshare struck it a glancing blow. Another time his small son picked up a stone to throw into the cattle pond, noticed how heavy it was and brought it to the house.

Varying amounts of metallic iron containing nickel are found mixed with the stony matter of most of the meteorites. More rarely they are composed of nickeliferous iron alone. After the Second World War, a man from Hutchinson, Kansas, brought down a mine detector and went over some of the fields looking for buried meteorites. The first day he excitedly dug up a tin can. But on a later day he made a real strike. Not far from the house, near the fence of the hog lot, his sensitive detecting apparatus revealed the presence of a 400-pound meteorite buried only two or three feet below the surface.

Almost all of the meteorites of the Kimberly farm were discovered on about eighty acres of the homestead. Only dust and small fragments have been picked up on the neighboring farms. Apparently, as frequently happens, a larger meteorite burst just before striking, raining the ground below. The great Meteor Crater of northern Arizona, which Nellie and I swung south to see when the summer trip had ended, illustrates the capacity for destruction of a really large meteorite—in this case believed to have weighed at least 12,000 tons. The explosive impact hurled out 400,000,000 tons of rock and excavated a pit that after 50,000 years of weathering is still four-fifths of a mile in diameter and 570 feet deep. Across the Kimberly fields the path of their exploding meteor extends for something like half a mile. That night as we watched the Perseids, Nellie and I discussed the amazing coincidence that placed the one person in the region capable of recognizing the importance of these stones from the sky on the very farm where they were concentrated.

Half a mile east of the farmhouse, where a field was now planted with corn, the Kimberlys had erected their original pioneer dwelling in the 1880's. We stopped for a long time here. During the night we returned to this spot again and again. Our feet left a maze of tracks in the dust that no doubt puzzled the first person to drive that way in the morning. All around the horizon we could see a low, diaphanous gray-white wall that seemed rising upward and thinning as it rose—the dust in miles of prairie air made visible by moonlight and starlight.

Above this wall of floating dust lifted the glitter of constellations and galaxies. As our eyes wandered among them, they brought to mind old Messier, the French comet hunter, and his paradoxical claim to fame. In his nightly search of the heavens, Messier listed 103 "nuisance spots," indistinct luminous areas in the sky that interfered with his work. In the course of time, improved telescopes revealed that these luminous spots were largely galaxies, island universes, the most exciting objects in the sky. All that later astronomers had to do to locate them in the heavens was to consult Messier's chart of nuisance spots. Even today many galaxies are listed by the designating numbers he gave them. Thus the great Andromeda Galaxy is referred to by astronomers as M 31—number 31 on Messier's list. This early scientist's lifework on comets is virtually forgotten today; his astronomical fame rests on his listing of nuisances he wished to avoid.

Standing there, facing the eastern sky, we noticed that our eyes grew progressively keener, training themselves as the night advanced to catch the fainter trails of light that winked out almost as soon as they began, the paths of little meteors too short-lived to provide even a beginning for making a "wish upon a falling star." Once in the space of no more than sixty seconds, four meteors streamed in different directions across the sky. One plunged almost vertically down into the northeast. Another traveled a long way, yellow-hued and almost horizontal. The highest of the meteors shone with a bluish cast; those lower down—seen through denser atmosphere nearer the earth—marked their transitory paths with yellowish brilliance. If the tracks of these shooting stars were left arrested on the sky, we thought, how streaked and crisscrossed would be the heavens when morning dawned! And if all the speeding celestial fragments we saw that night had come to earth in Kansas, how enhanced would be its fame as a mine of meteorites!

That fame, in recent times, has been due to a remarkable extent to the activity of one man, Dr. Harvey H. Nininger, of the American Meteorite Museum, at Sedonia, Arizona. With indefatigable enthusiasm, Dr. Nininger traveled up and down Kansas, speaking at farmers' meetings and country schoolhouses, arousing interest and exhibiting meteorites. At one time he was accounting for half the meteorite discoveries being made in the world. For several weeks he once hunted over the countryside near Haviland, staying with the Kimberlys. They spoke of him almost as one of the family.

Half a dozen times a year, on the average, meteorites are observed, somewhere in the world, in the very act of striking the earth. One weighing 150 pounds landed within the limits of Colby, Wisconsin, on the Fourth of July, in 1917. Another, on July 6, 1924, struck a highway near Johnstown, Colorado, just after a funeral procession had passed by, and its fall was observed by all those assembled at the cemetery only a few rods away. In Sylacauga, Alabama, on a September day in 1954, Mrs. Hewlett Hodges was resting on a couch when a ten-pound meteorite crashed through the roof of her house and struck her a glancing blow.

Because of the sudden increase in density of the air before them, meteors soon lose their planetary velocities in the atmosphere. They usually reach the earth at the speed of ordinary falling objects. One was once observed in Sweden striking ice only a few inches thick and bounding off without breaking through. Moreover, most of the intense heat of the meteor's flaming course through the upper atmosphere has been lost by the end of its slowed-down descent. In 1890, during a shower of meteors near Forest City, Iowa, one fragment fell on a stack of dry hay without setting it afire.

Far away, almost directly below Perseus, the few street lamps of Haviland winked in a small cluster of stars on the dark horizon line. Our car stood white and ghostly, its top and hood reflecting the glow and glimmer of the moonbeams. All around us the roads were deserted and, save for an occasional night insect singing along the way, wrapped in silence. The breeze flowed gently out of the south, bland and presaging the heat of the coming day. Once we caught the low call of a burrowing owl. Another time a killdeer went off in the moonlight with its wild plover cry. Our footfalls, silent as the moonlight, were cushioned by the carpet of dust. And as we moved about, changing our positions, our moon shadows moved with us, imprinted on the dust of the road and slowly elongating as their source descended toward the west.

Imperceptibly that night the great wheel of the constellations kept turning. The sword of Orion had long since been withdrawn from the haze along the eastern horizon. We had to look ever higher in the heavens to see Perseus. But hour after hour the meteors kept coming, those celestial sparks of running fire, mysterious and awesome in the night. It is easy to understand how their remains, objects descended from the void, the "black stars" or "thunderstones" of the ancients, were long valued and often held in superstitious reverence. More than 3,000 years ago, a Hittite king compiled a list of his treasures, noting, along with his gold and silver and bronze, "black iron of heaven from the sky." Early man, long before iron was smelted, was using the metal of meteorites in the manufacture of tools.

NIGHT OF FALLING STARS

Although meteor showers were recorded in China as early as 650 B.C., it has been only during very recent generations that they have received serious scientific consideration. Even as late as the eighteenth century the idea of the celestial origin of meteorites was ridiculed. The great French scientist, Lavoisier, maintained that they were ordinary stones that had been struck by lightning. A leader in attacking the "superstition" of falling meteorites was the French Academy of Science. Then, in 1803, thousands of stones fell on the village of L'Aigle, not far from Paris. The Academy sent a representative to investigate. He reported there was no question about it—they were meteorites and they had fallen from the sky. Then, as is the way of the world, the French Academy of Science hailed its representative as a great explorer who had discovered the existence of meteorites.

It was not until thirty years later that the real study of meteoric phenomena began. Its impetus was a night of falling stars unparalleled in modern history, the first and never-again-approached Leonid shower of November 12, 1833. It brought terror to many parts of the world. In Boston, a fifteen-minute count indicated the meteors were flashing across the sky at the rate of nearly 30,000 an hour. A reporter in Georgia wrote that it appeared as though "worlds upon worlds from the infinity of space were rushing like a whirlwind to our globe, the stars descending like a snowfall to the earth." Even the calendar of the Sioux Indians records this night of fire in the sky. Discussed for weeks, the prodigy stimulated widespread study of the problems connected with the life and death of these heavenly bodies.

We were sitting in the car, motionless and silent, sometime after three o'clock that morning, when I looked down at the side of the road. A pinched little face was peering up at me in the moonlight. Edging uncertainly past was a half-grown, gray and white farm kitten stealing by on some nocturnal hunting expedition. I slid slowly out of the car and we made friends. The kitten waved its tail and rolled in the dust. It rubbed its back against an extended pencil that bounced up and down along the hills and hollows of its protruding spine. Gaunt and apparently half starved, it appeared to have lost most of its voice, uttering only faint, one-syllable meows. Never once did it purr. It seemed not to know how. And in truth it probably had, in its small life, little to purr about.

The only food we had in the car was a box half full of graham crackers. I opened the trunk and got them out. When I offered a piece between thumb and forefinger, the kitten seemed confused. No one, apparently, had ever fed it by hand. I laid the cracker on the ground. It fell on it ravenously. At first it appeared to lack sufficient saliva for swallowing. But then it ate and ate and ate. Never again in all its life would it encounter this taste it was experiencing for the first time during this moonlit night's adventure. How many times would its little mind remember it? If kittens dream, how often, I wondered, would the haunting taste of graham crackers return in future catnaps?

We had been there for some time, our three shadows, two large and one tiny, dark upon the dust of the road, when I became aware of a fourth shadow. A jack rabbit had joined us. Not more than twenty feet away, it was watching us with curiosity rather than fear. I flashed the beam of a hand torch over it. Its eyes gleamed huge and glowing red. But it showed no alarm. For that small time we seemed transported back to an age that antedated fear, to some lost Eden we now but vaguely remember and rarely glimpse again. That moment of silence and moonlight, of strange and lonely companionship, seemed a moment in another world.

Then the spell was broken. The kitten saw the rabbit. It flattened in the dust, ears pricked forward. In this position it remained for a minute or more. Then, one foot slowly advancing before the other, it began the cautious stalking of a quarry fully four or five times its size. Unhurried, unalarmed,

NIGHT OF FALLING STARS

confident in its speed, the jack rabbit would bound a hop or two away, then sit down again to watch the stalking approach of its adversary, a very small kitten filled with courage, ambition and graham crackers. And thus they disappeared—their forms gradually losing distinctness, fading away like a dream into the summer night.

After 4:30 that morning, just before daybreak, the meteor shower began to diminish. The last falling star I remember seeing shone bluish, brilliant, going away from us toward the north. Perseus now had ascended far over our heads. All that night the skies had been as silent as the land. We were apparently off the main flight lanes. No single plane droned overhead. Slowly around us, in the first faint light of dawn, we saw the dusty roads emerging, the whole flat prairieland rising out of the night. Minute by minute the light strengthened and the stars paled.

It was after five when we passed the meteorite farm for a final time. The weather-beaten boards, the dusty yard, the dust-grayed trees now all stood out distinctly. In the dawning of that day there were no clouds, little red, only the hard glitter of the sun ascending with a disk of burnished brass. The night, with all its magic and memories, the night of the falling stars, was over.

TWENTY-SIX

THE GLASS MOUNTAINS

UNDER the prairie dawn, we took stock as we drove west to Greensburg for breakfast. We had gone without sleep for almost twenty-four hours. Yet we felt buoyant, elated by the beauty and mystery of that transcendent night with its lines of fire drawn across the sky.

I remember that after breakfast we peered into the cavernous depths of Greensburg's special attraction, what is said to be the deepest and largest well ever dug by hand. It was excavated in 1887 at an estimated cost of $45,000. The bottom of this titanic underground barrel, cased with stone and thirty-two feet in diameter, lay 109 feet below us.

By the time we had climbed into the car for the run south to Coldwater and east to Medicine Lodge, the burnished brass disk of the rising sun had become the white disk of the risen sun. Heat and our weariness mounted. There was a curious dreamlike quality, a misty, magic quality, about the hours that followed. The sharpness of remembrance waxes and wanes. It is as though fatigue, like drifting fog, from time to time partially obscured the scenes and events around us. Yet that morning was one of the most enjoyable possible. Everything seemed new and unfamiliar and filled with interest. We felt as though we could go on and on, continuing without stopping for days on end.

About 7:30, somewhere east of Coldwater, we came upon a large badger killed by a car in the night. A quarter of an hour later, hillsides stretched around us mantled in white where escaped snow-on-the-mountain was in full bloom. And by eight o'clock we had come to the valley of the butterflies. Here, where the highway dipped into an eighth of a mile of stream bottom, orange-red wings with irregular edges and a span of more than two inches fluttered and shone, weaving a pattern of color across the green foliage of the trees and lower vegetation. These brilliant goatweed butterflies, northern relatives of the dead-leaf lepidoptera of the tropics, seem to disappear the instant they alight. Their striking colors are replaced by the dull, subdued brownish hues of the underside of their wings. For some undiscovered reason these leaf wings had congregated in this small valley close to the Oklahoma line.

After the butterfly valley came the scissor-tailed flycatchers. With plumage that ranges from white to black through pearly gray and salmon pink and bright scarlet, with a forked tail that is almost twice the length of its body, this "Texas bird of paradise" is like no other species in North America. During their spectacular courtship flights, the males reach a climax of grace and airmanship, sometimes somersaulting downward, looping the loop time after time.

We stopped for a long while to watch a running battle between a Bewick's wren and one of the flycatchers. Darting among the bushes, almost constantly in motion in sudden starts and stops, the wren sought to reach a nest that—although the normal time for eggs is June, not August—a female flycatcher was defending. Her small brown-clad tormentor would dash toward the nest, then bolt away again with the flashing colors of the flycatcher only a foot or two behind. At least six times we saw it reach, or almost reach, its goal. But each time the scissor-tail rushed into action like a buzz saw, whirling downward in a twisting pursuit that drove the wren helter-skelter into the lower vegetation.

From the sagebrush, as we advanced and the heat of the day

increased, the clattering uproar of cicadas soared like the sound of a factory running full speed. A cicada song that was entirely different came from the treetops. It rose up the scale, higher and higher, increasing in intensity and volume, only to fall away in a descending, anticlimactic buzz or sizzle at the end. For miles now, plumed grass, with heads as soft as kittens' fur, ran in shining clusters along the road beside us.

In *A Guide to Bird Finding West of the Mississippi*, Olin Sewall Pettingill writes under the heading, MEDICINE LODGE: "Between 10 May and 10 September, one is almost certain to see a few Mississippi kites along the first ten or fifteen miles of U.S. Route 160 east or west of this town." We were on U.S. 160. We were riding toward Medicine Lodge. As we drove eastward we kept a constant watch for these black-tailed kites, slender and beautiful hawks that for years we had wanted to see.

Both at once we saw them. We saw not one but seven among the trees bordering a small stream. I pulled to an abrupt stop. We were twenty miles from Medicine Lodge. For half an hour we sat there watching the hawks at rest and sporting in the air. Their delicate shadings of gray, their silvery heads shining almost white in the sun, gave the adults a particularly striking and noble appearance. They held their heads high. And in the air they surpassed even the grace of the scissor-tailed flycatchers that had, but a few minutes before, impressed us so greatly. The slender-winged kites seemed the personification of beauty in motion. We watched them at play in the sky, diving and twisting in mock combat with one another. Superlative masters of the air, they veered, plunged, rocketed up again, their speed and grace and perfect control, their effortless abandon a glorious thing to see. In their sudden movements they seemed like the thrust and parry of rapiers in the air.

More than one of these birds appeared immature, spotted and streaked, light and blue-gray. They drooped their wings and called loudly whenever an adult landed nearby. And when they, themselves, swooped to alight on a branch their tails tipped high as though they had not quite mastered the art of judging their speed and in consequence had almost overshot the mark and somersaulted forward. In this region, after the one or two, rarely three, white or bluish-white eggs have been incubated for fully a month, the young kites hatch during the month of June. In these August days the new brood of the year was still perfecting the always difficult art of landing.

After the middle of the month, in August, it has been noted, these birds increase their activity in feeding. They thus put on fat, reserve resources of energy, before moving south in September. In the main, their food consists of the larger insects, the noisy cicadas and the numerous grasshoppers. As many as twenty kites have been seen circling and diving about a man on horseback as he rode through brushy or grassy country, swooping to seize the locusts and cicadas scared up by his advance. The captured insects are held in the talons of the hawks while the inedible parts are stripped away. Then the soft body is eaten on the wing. The kites sometimes demonstrate their astonishing powers of vision by plunging down from a height of 300 feet to pick a grasshopper from a leaf.

Insect fare is varied from time to time by the addition of a mouse, a toad, a small snake, a frog or lizard. Among the branches of the streamside trees or flying over the open fields nearby were a number of small birds. They gave no sign of fear and paid scant attention to the kites sporting in the air above them. Apparently these fleet-winged hawks never prey on other birds. In fact, although it is courageous in the defense of its nest—at such times attacking even a man—the Mississippi kite, under ordinary circumstances, is described as peaceful and inoffensive—the gentle hawk.

Arthur Cleveland Bent begins his account of this species in his *Life Histories of North American Birds of Prey* with the statement: "I have never seen this kite in life." We now had seen it "in life." And what a wonderful, swift-ranging life it is! A few miles farther down the road we came upon a group of five more, and not far beyond that eight others sweeping and tilting a hundred feet or so above a wide expanse of cultivated fields. Watching them through our glasses, we were fascinated by the mobile shifting of their tails. They spread, tilted, closed, rose on the sides, keeled in the middle. They were expressive of stresses and strains—even, it seemed, of the emotions of the flying hawks. Continually as the birds shifted in the play of the breeze their tails were varied in size and shape and position. The kites circled. They halted momentarily in the air to search the ground below. Slowly they moved away across the fields until they became small in the distance. In all, before we left the vicinity of Medicine Lodge that morning, our count of these beautiful and graceful birds was twenty-seven.

The road that carried us southward into Oklahoma wound endlessly, turning with the twisting of the Medicine Lodge River. The fields around us now were copper-colored. The sky above shimmered with the white incandescence of the sun. The air, dry and scorching, rushed in the open windows of the car like the breath of a blast furnace. A few days before I had broken our thermometer, and we could only guess at the temperature. We came to Alva and decided to drive on to Cherokee to be close to the Great Salt Plains the next morning. Except for a single roadside catnap of less than fifteen minutes, we had now been awake more than thirty hours, and the heat and fatigue bore down with ever-increasing weight. Cherokee was a sizzling griddle. Sparrows perched disconsolately with bills wide open. They seemed distressed. So were we. Our bills were wide open, too.

Nowhere in this burning town could we find a place to stay. But we did discover a small cafe where electric fans buzzed as impotently as flies in an oven but where the food was good in spite of the sweltering heat. As we were turning back

WILLOW on the banks of the Arkansas River near Wichita.

MUD FLATS, cracked in the summer drought, near the Great Salt Plains of Oklahoma. Below, massed cactus.

toward Alva we passed a large thermometer hanging in the shade on the front of a store. I backed up. We peered at it twice. The mercury stood at 110 degrees F.

Slowly we crawled across the torrid landscape, retracing the way to Alva. We passed dry ditches, their bottoms dotted with crayfish towers. We came upon a hot and sluggish vulture loath to leave the carcass of its roadside rabbit. We saw a hundred dust devils wandering over the burning surface of the parched land. Across every field that had been plowed, these whirling funnels raced and spun.

A little before two that afternoon we were back at Alva. There we treated ourselves to the best air-conditioned motel. For more than thirty-three hours we had been awake—under the night sky, under the fiery sun, in the greatest heat we were to encounter on our trip. We had been exhilarated by all we had seen—falling stars, kites and flycatchers and orange-red butterflies. But now the stimulation was over. Our feet dragged. We were borne down by the heat, bleary-eyed from loss of sleep. With the automobile under the shade of a carport, we took cool showers and dropped into sleep like spent rockets. About eight o'clock that evening we roused ourselves long enough for a light supper of bacon-and-tomato sandwiches and frosted root beer. The thermometer at 8:30 still stood close to 100 degrees. Back in our room we fell once more into fatigue-drugged slumber. So ended the day that followed the night, the night of the falling stars.

Paradoxically, it was the next morning, after more than fourteen hours of sleep, that we felt most overpowered by the apathy of fatigue. Our interest flagged. Our minds slowly shook off the drug of slumber. We seemed to take all morning to awaken. Dull-witted, we rode across the Great Salt Plains east of Cherokee and north of Jet. Always I want to see things through my *own* eyes rather than through the eyes of others; rarely is any place exactly as I had expected to find it. So it was with the salt plains. Here we found more than the expected shimmering world of crusted white. Here we found, in addition to the 10,000 acres of salt flats, 10,000 acres of impounded waters and 12,000 acres of upland and forest. This is a place we want to see again; for we saw it only half alive. Even in the apathy of those first hours of the morning, we regretted each moment of lessened awareness. For life in fascinating forms was all around us, even on the white prairies of the salt.

Here that bird of far-southern ocean beaches, the snowy plover, is a nesting species. We saw three of these shore birds, each smaller than a piping plover, alight not far from the Salt Fork of the Arkansas River. So white is their plumage that the instant they became motionless they seemed to disappear, leaving only their small black shoulder marks visible—like the smile that remained after the Cheshire Cat faded away in Alice's Wonderland.

Among the trees in the shaded woodland stretches, activity ranged through all the branches. Whole families of western kingbirds conversed in squawling twitters far above the Bewick's wrens that flitted through the underbrush. Once we passed an ancient locust tree, its dead top filled with roosting turkey vultures. Again we came upon a nest that shone out like a bright, yellow-hued flower. It was formed largely of strands of parasitic dodder that grew nearby.

Still tired that afternoon, but with minds once more awake, we crossed the Cimarron River thirty miles below the Great Salt Plains. At places along this portion of its course the sandy stream is a full mile wide. Now all the visible water ran in a narrow creek along one side of its shallow bed. Like most western streams in August, the Cimarron was flowing as an upside-down river. More of its water moved below its bed than above it; more was traveling underground than at the surface.

Low in the west, as we rode south, we could catch the outline of sandhills created from the dry river bed in times of

THE GLASS MOUNTAINS 271

wind and drought. Also to the west, as we neared the little town of Orienta below the Cimarron, we glimpsed another range of hills extending in a dark-red line along the horizon. We turned toward them. These were the strange Glass Mountains of north-central Oklahoma. Out of the shimmering distance the red slopes drew nearer. Glinting dots of brilliance spangled them from top to bottom. They seemed coated with the fragments of ten thousand broken mirrors, each catching and reflecting the sun.

To geologists, the Glass Mountains of Oklahoma represent "an outlier of the Blaine Escarpment"—that great gypsum formation that extends over much of the western part of the state. It forms, roughly, a triangle, its apex at the Kansas border, its base above the Wichita Mountains. In industry and commerce the uses of gypsum are innumerable. It forms the white chalk of schoolrooms. It is widely employed as wallboard in construction. Plaster of Paris is gypsum, its name being derived from the fact that it was first produced from raw material mined at Montmartre in the French capital. In North America, sedimentary rock of this material, produced as a saline residue, is found from Nova Scotia to Arizona. One form, known as selenite, occurs in layers of clay as sheetlike crystals. It is selenite in broken sheets that glitters all across the flanks of these Oklahoma hills. Derived from the name of the Greek goddess of the moon, Selene, the word "selenite" reflects an odd early belief. Men noticed that the plates of this mineral reflected more light when the moon was full than when it was partially obscured. From this they jumped to the erroneous conclusion that the mineral, itself, waxed and waned with the moon.

When we reached and wandered among them, we found ourselves in a raw, red world. The eroded buttes were red. The dust was red. The sweat we wiped from our faces was tinged with red. Only the dull green of a rare yucca and the glittering pieces of gypsum relieved the monotony. We picked up plates of selenite like fragments of broken window panes. We seemed walking among vast dump heaps where red earth and broken glass were piled together.

There are places among these mountains where blocks of solid gypsum, as much as six feet thick, have been shaped into bizarre formations. These outcroppings resemble human faces or domes or minarets rearing above the red slopes. One such formation so strongly suggests the towers and mullioned windows of a medieval church that the whole butte is known as Cathedral Mountain. But here, at the far eastern edge of the escarpment, it was the fragments of glasslike selenite that caught and held the eye. More resistant to erosion, it had endured, outlasting the softer shales and clays.

Lying on the page of an open book beside me as I set down these words is a fragment of this mineral that I picked up that day. It is about half the size of my outspread hand. Although it is a quarter of an inch thick, I see that the printing beneath it is clearly visible. I can read it as through a pane of window glass. Held to the light, the edges of this plate reveal fine parallel lines. I run a thumbnail along an outer line and, with almost no effort, split away a sheet of the mineral, tissue-paper thin. I rake my thumbnail across the flat upper surface of the plate. A white scratch appears behind it. Selenite is soft, as minerals go. In spite of this fact, in spite of the innumerable thin cleavage plates which it possesses, it has been able to endure long after the material that originally surrounded it has been eroded away. The sheathing of selenite, spreading in a glittering coat of mail over the butte sides, imparts a special, almost legendary character to these shining mountains.

Turning back toward Kansas, amid the endless clatter of the range-land cicadas, we bore to the east. Through Pondcreek, toward Caldwell, we followed the line of the old Chisholm Trail. In the early days, tens of thousands of bawling longhorns had plodded northward along this same path heading toward the railhead at Wichita.

TWENTY-SEVEN

ON THE RIM OF THE WORLD

IN the days that followed we drove north and west and south again. Our path scrawled a great inverted "U" on the map of western Kansas. Like a broad teeter-totter, this rectangular state tilts upward toward the west. The traveler who drives the 400 miles from the Missouri line to the Colorado border ascends 3,000 feet. On the last third of his journey he finds himself in an area of high plains, flat and windy, vastly different from the rolling eastern portion of the state. A land without hills, almost without trees, it extends on and on toward the west. It becomes drier and flatter until it merges at last with the dry, flat, high prairieland of eastern Colorado.

Above the great bend of the Arkansas, as we drove north from Wichita, above Dry Walnut Creek and Wet Walnut Creek, Blood Creek and Cow Creek, above Hoisington and the wide wetland of the Cheyenne Bottoms, we found ourselves in a curious country of stone fenceposts. We had left the trees behind now. Mile after mile our highway was bordered on either side by barbed-wire fences supported by rows of yellow or buff-colored posts of stone. Once we passed a fencepost quarry in a field. Slender rectangles of chalky limestone were outlined on the surface of an outcropping of rock less than a foot in diameter. Posts of stone were stacked at one side. They appeared much whiter than those in the fences. A characteristic of the Greenhorn or fencepost limestone of north-central Kansas is that it comes from the ground chalk-white or cream-colored but gradually weathers to a buff or brownish hue. It is also soft when quarried, making it easy to cut, while it becomes harder and stronger after long exposure to the air. Laid down beneath prehistoric seas, the chalky limestone appears at the surface in rather thin ledges, a foot or less in diameter, interbedded with shale.

For thirty or forty miles in this treeless land, as we drove north, we saw all the fields fenced with such posts of stone. In many places on the Great Plains trees are rare or absent altogether. But Kansas is the only state I have encountered that has made a census—even an estimated census—of the number of its tree inhabitants. For its 86,276 square miles, the population of trees has been set down as about 225,000,000. Most of them are rooted east of the Great Bend of the Arkansas.

How many are Osage orange trees nobody knows. But the number must be high. All across the Middle West and southward, in the days before barbed wire, hedges of these thick, thorny trees were employed to mark boundary lines and fence in livestock. At one time, crews of men went through the country planting Osage orange trees for settlers at from fifty to seventy-five cents a rod. We saw miles of them still in use in southern Kansas. Already the pale-green balls of their orange-shaped fruit were fully formed.

Long before the settlers came and the first hedge was planted, the fame of these trees was so widespread that Indians traveled hundreds of miles to obtain the wood. Twice as hard as white oak, two and a half times as strong, the heartwood was greatly prized for making bows. Even today it is considered by many archers to be superior to yew. A century and a half ago, Indians would trade as much as a horse and a blanket for an exceptionally good Osage-orange bow. From the hard, resistant wood, the pioneers fashioned many things—axles, pulleys, tool handles. By boiling the chips they produced, as the Indians had done before them, a beautiful golden or orange-

ON THE RIM OF THE WORLD 275

yellow dye. At one time, blocks of the long-wearing wood paved the main streets of a number of American cities.

To the French explorers who came down the Mississippi, the tree was the *bois d'arc*, or bow wood. Another of its numerous names, referring to the globular fruit, is hedge apple. Within the spongy mass of these balls, with their milky juice, the seeds are so minute that it requires more than 12,000 to weigh a pound. Quail and fox squirrels are among the very few creatures that turn to this source of infinitesimal food.

Originally the range of these trees extended largely through the territory of the Osage Indians, from southern Missouri to northern Texas. Since then the Osage orange has been introduced successfully throughout much of the United States. It is resistant to heat and drought. It thrives under many conditions. Why, then, was it restricted to a relatively small range? Why, of its own accord, had it not spread widely across the country?

We puzzled over this mystery as we continued north. And as we went we saw another of those innumerable riddles of distribution taking shape around us. For a time scissor-tailed flycatchers perched on the fences along the way. Then we saw them grow rare and disappear entirely. We had passed beyond their local range. Yet the country seemed no different in any essential way. Driving south on a later day, 150 miles to the west, we suddenly found ourselves once more among these beautiful flycatchers. We had crossed an invisible boundary. We had entered again the territory of the scissor-tails. So on a morning in Montana we had found ourselves almost suddenly amid lark buntings as we crossed the boundary line of their local distribution from the north.

What determines these more or less stable frontiers of distribution year after year? Sometimes it is food. Sometimes it is a question of nesting opportunities. Again, at the fringes there may be an insufficient density of population to provide mates or a replacement if a mate is killed. In some cases the competition of another species with an adjoining range may counterbalance the pressure of the species outward from its own center of population and thus produce a relatively stable boundary at the limit of its range. For all forms of life, the mystery of distribution is one that has absorbed scientists for generations. Occasionally the answer is simple, as in the case of the fern that turns up only where there are outcroppings of limestone. But most of the time a whole combination of complex factors, often obscure, is involved.

Across all the rolling land above Salt Creek and Saline River, every dip roared with the sound of innumerable cicadas among the browning sunflowers. Somewhere in this region we noticed a wall made entirely of stones as round as cannonballs. And not long afterwards we encountered miniature hot-dogs, advertised at a roadside stand as "Pocket Pups." Everywhere, interminably, on these Kansas summer days, the grasshoppers promenaded across the hot pavement of the highway. We saw each approaching car surrounded by a shining cloud of locusts rising from the road before it. And whenever we stopped at a filling station we found English sparrows waiting to dine on the dead grasshoppers falling from cars or brushed from radiators.

Thus the feeding of these sparrows had completed a full circle. Originally, both in England and America, it was horses and stables that provided them with their main source of nourishment. When the first automobiles appeared in England, and the rivalry between chauffeurs and coachmen was at its height, it was the habit of the latter to shout after each passing car: "Sparrow starvers!" Now, in another land, even in the West, that last stronghold of the horse, we were seeing these same birds during these summer days turning not to horses but to automobiles for their food.

Once, as we were going down a long, straight stretch of road, I put out my hand to feel the temperature of the air. A flying grasshopper struck my palm like a beeliner from the bat of a world series ball player. At first I thought it had gone straight through like a bullet. For a long time afterwards my hand stung and hurt. I could well imagine the sensations of a Dalmatian dog of which I had heard in Wichita. During a heat wave in July it was riding toward Haviland with its head thrust out of the car window to get the breeze. Suddenly it let out a yelp, jerked its head inside and howled with pain. A grasshopper had hit it squarely on the nose.

We were near the top of Kansas when we turned west. During the summer months, any road in this Hub State of the Union is a sunflower trail. Everywhere we saw the golden disks. Calico ponies fed among them on the rangeland and beside the road the massed stems ran like hedges along the fence rows. Riding down highways lined with sunflowers is like riding in a parade with the faces of spectators dense on either hand. The peering faces of the sunflowers were with us all the way. Already in the upland fields many of the heads were dry, the seeds that would nourish manifold forms of life close-packed in spiral lines within them.

As we went west, deeper into the rain shadow of the Rockies, farther into the higher, drier land of the state, the sunflowers shrank in size. The annual rainfall at the eastern border of Kansas is about forty inches. At the western boundary it is only about fifteen. This swift decrease in moisture is reflected in the shorter stems, the smaller faces of the western sunflowers. At first the yellow disks rose above the fenceposts. Progressively they became more and more drought-stunted. Their disks contracted, their stems grew dwarfed. At last, in the driest stretches, they were less than knee-high. We saw their forlorn little faces peering at us just above the lowest wire of the fences.

From Wakeeney, west through Grainfield and on to Oakley, our road paralleled the flow of the Saline River. The land now was almost as flat as a waveless, windless sea. Only the faint swells gave a slight up-and-down play to our advance. The open fields stretched away as big as townships. Each farmhouse, surrounded by its buildings and shaded by a few planted trees, stood out island-like on the level land. Viewed from afar, every small community seemed formed of skyscrapers. These structures were those skyscrapers of the plains, towering white grain elevators, clustering together like the many-celled nests of some giant mason wasp. As we neared Oakley, the day's journey's end, in the heat distortion of late afternoon, all the distant farmhouse and elevator clusters appeared disconnected from the earth, lifted up, floating on a shining lake of mirage water, islands in the sky.

The next day, the long, dry road that runs from Oakley past Shallow Water to Garden City carried us south. Horned larks, gray as the dust, flitted up from the roadside and shrikes, perched on fenceposts, fed on captured grasshoppers. In great grasshopper years, heaps and windrows of discarded legs and wings will build up around the favorite feeding posts of the shrikes. Once a prairie falcon swept by in swift, sure flight. Again we saw a pectoral sandpiper circling overhead. What was a "grass snipe" doing in this dry, flat land? It knew best. For only a short distance down the road we came to a shallow pond surrounded by an acre or more of pink smartweed in bloom.

All the highways on this level land ran straight, roads without turnings. The country for a million acres around us seemed to have been rolled until it was perfectly flat. In this uniform, almost featureless region, the scenery was mainly scenery of the sky. We saw cloudscapes rather than landscapes. In the clear, dry air the low horizon appeared immeasurably remote. We had the impression that only the curvature of the earth prevented us from seeing on and on to infinity. We seemed riding on the rim of the world.

The wind rose as we advanced. Fields became smoky with dust. That probably is the story of every morning on these high plains. More than once as we rode south we came to an open field that seemed on fire, the fine dust rolling up as the wind gained strength. Each time we would ride through a gritty cloud and then come out into the sunshine again. We had encountered a one-field dust storm—but what a field!

According to the U.S. Weather Bureau, the western third of Kansas ranks as one of the windiest inland areas in the United States. Here there is little vegetation to slow down the moving air. Tests have shown that a breeze blowing nine miles an hour three inches above the prairie vegetation is reduced to three and seven-tenths miles an hour where it comes in contact with the tops of the plants and to one-tenth of a mile an hour at half that height. In this part of the state, violent hailstorms are a feature of the windy prairie winters. Some thirty miles north of Oakley, at Selden, Kansas, a 2½-hour hailstorm in the spring of 1959 buried the area under eighteen inches of ice. Property damage in the town was placed at nearly a quarter of a million dollars. Over the years statistics have revealed that the damage from hail is five times greater in western Kansas than it is in the eastern portions of the state.

In the wind that morning, and under the hot sun, we saw dust devils endlessly forming across the flat expanse. They went spinning away until each in turn, with strength spent, dissolved and disappeared. Some of these toy twisters whirled their funnels of dust a hundred feet or more into the air. One attained a height of fully 500 feet. It is a rough rule of thumb of meteorology that the life of a dust devil can be calculated by its height, a thousand feet equalling an hour. Thus the 500-foot funnel had a life expectancy of half an hour. Always the largest of the air-borne particles are carried at the outside of the funnel. There the speed, and hence the carrying capacity, of the wind is greatest.

Once a long truck cut around us and raced away down the road, leaving behind a series of baby twisters spinning and carrying the cut roadside grass aloft. The atmosphere of this land of tornadoes seemed set on a hair trigger, less stable than the more weighted air of humid places. Or so it appeared to us as we watched the birth and death of whirlwinds all along the way that day. Yet, in truth, it is the more humid eastern portions of Kansas, rather than the drier western areas, that experience the greatest number of destructive tornadoes.

A tornado is not, as is often assumed, merely an overgrown dust devil. The dust devil begins at the earth's surface as a whirlpool of air and builds upward. A tornado, on the other hand, is a funnel of whirling air that descends toward the earth. It begins revolving below heavy storm clouds. The dust devil is a product of sunshine, convection currents and the heat of the day. Tornadoes occur at night as well as during the day. They are often accompanied by rain, by hail that may hurl down ice stones as large as baseballs, and sometimes by tremendous displays of lightning in which the flashes may be green, yellow or blue. Dust devils range from less than ten to about 100 feet in diameter, while the funnel of the average tornado measures about 250 yards across. The direction of rotation of a dust devil is accidental. Some spin clockwise, others counterclockwise. But every tornado in the Northern Hemisphere whirls in the same way, counterclockwise.

Nearly 900 of these deadly funnels spun in the United States in a single recent year. They are born when a warm, moist wind collides with a dry, cold one. Exactly what occurs is still being debated by scientists. But the results are not debatable. As the Kansas saying has it: "You can't argue with a cyclone!" The average speed of advance of American tornadoes is about forty-five miles an hour. This, however, may be increased to more than a mile a minute, or the whole roaring funnel may come to a complete standstill for as long as twenty minutes. It is only when the lower tip of the funnel is in contact with the ground that the tornado makes its long sweep of destruction.

Deadly as cyclones are, some scientists are coming to believe that they may be, in the over-all picture, beneficial. They may provide a safety valve for pent-up energy. According to Dr. Theodore Theodorsen, of the University of Maryland, with-

ON THE RIM OF THE WORLD

out the type of air turbulence that produces a tornado, hundred-mile-an-hour winds might sweep across the face of the earth continually. Man would exist in a constant hurricane. Life would be almost impossible. The super-turbulences tend to sap the strength of the main flow of air and thus slow it down.

At Garden City we turned east to follow the Arkansas River once more. Here the stream was a small summer brook meandering down the widespread channel of its spring floodtime. We crossed the hundredth meridian and drove toward Dodge City across the plains where Zebulon M. Pike, in the fall of 1806, came upon a herd of buffalo that extended as far as he could see. "I believe," he wrote in his journal at the time, "that there are buffalo, elk and deer sufficient on the banks of the Arkansas alone, if used without waste, to feed all the savages in the U.S. territory one century."

Leaving the river at Dodge City we turned southward again in overpowering heat. Blazing days, and winds with no coolness in them, and nights when the thermometer is slow to drop —these, too, are part of nature's summer. Such conditions are normal for innumerable creatures, innumerable rooted forms of life. The seashore, the mountains, the cool north woods, the area of the summer resort, these represent but a small fraction of the world at this season of the year. On our journey we were seeking to *see* summer, not escape from summer, and these hours of soaring heat and dazzling sun represented part of what we set out to see.

For weeks here no drop of rain had fallen. This was the usual condition at this time of year. All this high, dry land lies deep in the rain shadow of the Rockies. During the era of homesteading in this region, fast-talking promoters made current the saying: "Rain follows the plow." That catchy falsehood brought scant comfort during the droughts of later Augusts. Bizarre and often irrational schemes were tried in earlier days to increase the rainfall in these arid sections of the West. One of the superstitious rites of the time apparently persists into the present. We saw numerous dead snakes hung belly-up on the fences. An old folk belief had it that a snake left in such a position would bring rain.

Probably the misconception that lived longest and died hardest was the idea that loud noises would shake raindrops from the sky. In the year 1872, the United States Congress was petitioned to send a hundred cannons to an arid spot and fire them all at once to test the theory. By 1891 public pressure became so great that an experiment of the kind was actually carried out near Midland, Texas. Salvo after salvo was fired into the sky. At the same time balloons with lighted fuses were released to explode high in the air. One detonated in the very heart of a dark cloud. But no rain fell.

It was in the region of western Kansas that the rain-makers of another generation reached their heyday. These self-styled "weather wizards" journeyed back and forth across the dry country. After contracting to produce rain in an area for a sizable fee—not to be paid unless rainfall occurred—the wizard would retire into a windowless shack. Wisps of exotically colored chemical fumes would appear from its chimney. They were supposed to mingle with the atmosphere and produce a shower. If, in the course of events, rain did happen to fall within a certain time after his effort, the rainmaker collected his fee. If it did not, he had lost little more than his time.

At one period the public faith in rain-making was so great that one railroad set aside a special "laboratory car" and sent a celebrated weather wizard on a grand tour of the sidings of the line, stopping at each and spending several days in his chemical hocus-pocus. In the main, the procedure seems to have been to pour sulphuric acid over zinc to produce hydrogen. This lighter-than-air gas, together with some coloring agent to give it a mystifying hue, billowed into the sky. It was supposed to combine with oxygen to form H_2O and to fall to the earth as rain. As modern research has shown, in arid regions where there is insufficient moisture in the air, not even present-day seeding of the clouds will cause a rainfall.

Somewhere below Meade that day we crossed the Cimarron riverbottom. Here our road was densely lined with sunflowers, their stalks interlaced with dodder. From one yellow border to the other a lizard, bright yellow itself, raced across the burning pavement. Beyond the Cimarron we came to Liberal, our last town in Kansas. In Nebraska we had passed through a community labeled "The Egg Center of the Nation"; in North Dakota another styled "The Artesian Well Center of the United States." Here at Liberal we came to a sign proclaiming it "The Pancake Hub of the Universe."

It was midafternoon when we came to the Oklahoma line. The Glass Mountains, the snowy plover, the Great Salt Plains lay 140 miles to the east. Before us stretched that narrow corridor of land, 150 miles long and less than 40 wide—the Panhandle of Oklahoma. Behind us now spread all the state of Kansas, our third home, the third commonwealth in which we had lived. Whenever we thought of it again, and remembered this latest visit to it, many things would come to mind: the beauty of the scissor-tailed flycatchers, the grace of the Mississippi kites, the joy of the little toads in the great storm along the Arkansas. But most of all we would remember that night of the falling stars—the softness of the air, the moonlight, the emotions of the hour, the silent, outspread land, and above it all the ethereal beauty of the sky crossed by those lines of moving fire. It had, for us, been one of our own Arabian Nights, of which there are but a few, a very few, in a lifetime.

TWENTY-EIGHT

PARADE OF THE DUSTY TURTLES

FOR more than 160 miles, the Oklahoma Panhandle thrusts its narrow strip westward to the boundary of New Mexico. It is sandwiched between Texas on the south and Kansas and Colorado on the north. A single highway, running for a considerable part of the distance like a median line, extends down the length of this rising corridor of land. It was along this road, that day, that we came to the beginning of an unexpected adventure.

The country was high, open range and farming land. An occasional stunted cottonwood rose against the cloudless sky, its top, killed by some previous drought, silvery and wind-polished and filled with bird nests. In this almost treeless area, the lack of suitable nesting sites is a principal limiting factor for a number of species. All the far-spaced trees we saw were bird apartment houses, their bare upper branches supporting, as though clad in bunched foliage, the stick-masses of many nests.

More than once beside the road, long stretches of rangeland were swept with a luminous silvery shine. The rays of the morning sun were catching the woolly stems and leaves of the plume *Eriogonum*. Each plant seemed clad in glowing hoar frost. Contrasting with this beauty of the morning, intermittent patches of dodder blighted the roadside, cobwebby, unpleasant, nightmarish, everything dead or dying under its

MOUNTAINS IN FLOWER, upper left. Right, Alpine goldflower. Lower left, yellow paintbrush, right, bistort.

ARCTIC gentians, top left. Right, rose or queen's crown. Below on the left, columbine, right, rocks on the tundra.

FLORISSANT shale bed in Colorado. Below left, leaf of a prehistoric water elm. At right, a fossil cranefly.

PARADE OF DUSTY TURTLES

touch, covering the ground beneath it like scum at a dry pond edge.

In the range country of southwestern Kansas we had first noticed a vine with striking triangular leaves upraised like wings. Here, beside this Oklahoma road, we pulled up and examined one closely. Oval fruit, green with longitudinal stripes of white, suggesting a miniature watermelon, had formed along the stem. This was the buffalo gourd, the chili coyote of the southern Great Plains. Giving a name, as Thomas Carlyle once noted in his journal, is indeed a poetic art. That art frequently has been exercised in the creation of picturesque names for western plants: woolly knees, cowboy's delight, fairy dusters, miner's candles and chili coyote.

It was along this road of interesting plants, not far from the chili coyote vine, that we came to the beginning of the turtle parade.

A movement at the roadside caught our eye. Out from among stunted sunflowers a dry-country tortoise pushed its way. From then on, west to Boise City—near the Black Mesa and the old route of the Sante Fe Trail—and south on the long, straight road that runs without meeting a community or hardly passing a house to Dalhart, in Texas, we were in the midst of a movement of turtles. We saw them plodding across the road ahead of us. We saw them appearing at the edge of flat fields. We saw them crawling from under the dry hedgerows of old tumbleweeds anchored against the fences. We saw a number killed by passing cars. Along half a hundred miles of highway that morning we encountered turles on the move.

What was happening? Where were they all going? Were we witnessing a migration or a time of abnormal abundance? All the wanderers we saw looked the same. They were all box turtles, arid-land and dusty turtles. Whenever we pulled off the road to examine one closely, we found it was about four or five inches long. Its general color was brownish. But it was handsomely decorated with yellowish lines radiating downward from three centers on either side of its shell. We hunted through our box of field guides, resting the books on a fender of the car. Nearly a century before, Louis Agassiz, at Harvard, had studied one of these creatures and had bestowed upon it the name by which it is known today: *Terrapene ornata*—the ornate box turtle. A native of the open prairie, partial to sandy, even semi-arid regions, it ranges westward to the skirts of the Rockies. On occasion it is found in the foothills as high as 6,000 feet above sea level.

On this late-August morning we were not in the midst of a turtle migration. Nor had we arrived at a time of abnormal abundance. We were encountering merely the normal activity of an extraordinary member of the turtle tribe. A native of dust-bowl areas, able to thrive under drought-land conditions, this species is famed for what Clifford H. Pope, in his *Turtles of the United States and Canada*, speaks of as its "astonishing abundance." F. W. Cragin, in pioneer times, reported in the *Transactions of the Kansas Academy of Science* for 1885 that these turtles were sometimes so numerous in the southern part of that state as to be a nuisance and "a cumberer of the ground."

In the very Oklahoma county through which we were riding, two scientists once collected 150—and they could have obtained more—during a single trip afield. Between Vernon and Wichita Falls, in northern Texas, the southern herpetologist John K. Strecker observed along the road forty-six ornate box turtles in forty-two miles. And not far from Hays, in the Smoky Hill River section of central Kansas, L. A. Brennan reported an average of one turtle a minute during a half-hour's walk along the edge of a melon field.

It is during the morning hours that these creatures are most in evidence. They are early risers. Just after the sun has cleared the horizon, they are usually found resting on the east side of tufts of grass or clumps of sage. They are taking their first sun bath of the day. During the hot hours of the prairie noon, they seek the shelter of bushes or retire into little pits they excavate in the ground. When we encountered them, they were wandering about in the long breakfast of the morning.

In a large part their ability to survive and multiply in sparse surroundings is the result of an omnivorous appetite. In fields where cantaloupe melons are grown, they sometimes cause considerable damage by biting holes in the ends of the ripening fruit. But this destruction is more than balanced by the long list of insect pests they consume. They devour grasshoppers, beetles and moth larvae in great numbers. One turtle was observed eating a small horned toad and another a slender lizard, an adult six-lined race runner.

Ornate box turtles one summer came regularly to feed on tomato worms in a field near Apache, Arizona. They appeared, plodding slowly down the rows, at almost exactly the same time each day. This methodical regularity in their habits has often been noticed. John K. Strecker tells of a trio of these turtles that he liberated in his back yard. They established themselves under a coal shed and emerged once each summer day to feed on little strips of beef. This meal was always served at six o'clock in the evening, and the turtles soon learned to appear at that hour. They timed their visits as though they possessed built-in alarm clocks. Animal psychologists who have studied these prairie turtles have found that they are unusually adept at learning to find their way through laboratory mazes. Among the turtles tested, they rank high in intelligence.

Oftentimes, when Strecker delayed his feeding, he would observe the trio in his backyard stand on their hind legs and scratch at his shoes and the bottom of his trousers in an effort to attract his attention. One boy who had fed a pet ornate box turtle on angleworms told me that frequently it would stand on its hind legs, balance itself in the air and shoot up its neck to snatch a worm that was dangled above it. Mild-tempered and docile, these turtles can make surprisingly fast movements in capturing active prey.

An astonishing instance of the kind is reported by A. I. Ortenburger and Beryl Freeman in their report of a University of Oklahoma Museum of Zoology expedition to the Panhandle region in June and July, 1926. "Some of the queerest sights seen," they note in Volume II of the University's Biological Survey, "were the attempts, often successful, of these supposedly slow-moving animals to catch grasshoppers. On one occasion, one turtle was seen to catch a large lubber grasshopper 'on the wing' by stretching the neck and literally jumping at the flying insect."

These "leaping turtles" of the Ortenburger and Freeman report are full of surprises. We gazed at them with added interest as they plodded deliberately beside the road. When we stopped and approached closely, in the immemorial manner of their kind they pulled in their legs and heads, curled in their stubby tails—tails that are shorter in the females than in the males—and folded up the hinged ends of the lower shell, shutting themselves up tightly within their horny boxes. The cracks in the lower part of this armor were too fine for claws to enter and obtain a purchase. They were so narrow, I found, that they would hardly admit the edge of a sheet of notepaper. Tight and strong, these shells are proof against almost all of the turtle's foes. But they are no protection against the most deadly enemy of all—the speeding automobile. Large numbers are killed, I was told, after every summer thunderstorm when the rain brings out the turtles all along the highways.

Rain they like, but ponds and pools they avoid. If one is placed in a puddle it scrambles out in haste. In rangeland, they sometimes fall into water tanks and drown. This makes all the more remarkable an observation related to me by Dr. Charles M. Bogert, head of the Department of Herpetology at the American Museum of Natural History. While conducting experiments at that institution's Southwest research center, near Portal, Arizona, he kept ornate box turtles and horned toads in the same pen. To provide a supply of food for the lizards, he dumped ants into the enclosure from time to time. They tended to gravitate under rocks in the pen where the turtles were hidden. There they swarmed over the motionless animals, annoying and biting them. More than once Dr. Bogert saw these ant-infested turtles perform the same remarkable rite.

In the center of the pen an ancient grinding stone, depressed in the middle, held a pool of water several inches deep. Bearing its load of clinging ants, one of the turtles would walk to this pool and enter headfirst, going only about half way under. When the submerged ants on the fore part of its body had floated to the surface, it would scramble out. Then it would reverse its position and back down into the water to repeat the process, remaining half submerged until the rear portion of its body was similarly cleared of the unwanted insects. Dr. Bogert watched this performance closely and saw it repeated on several different days.

Another noted American herpetologist, the late Dr. Karl P. Schmidt of the Chicago Natural History Museum, was walking down busy Michigan Avenue one November day in 1930, when he was amazed to see hundreds of baby ornate box turtles crawling over the grass at the edge of Grant Park. How they got there, on the main thoroughfare in Chicago, is a mystery to this day. It is speculated that a temporary craze of turtle racing had run its course and some promoter had unloaded his stable of small turtles by dumping them during the night at the edge of the park. Dr. Schmidt and an assistant collected more than 500 which they later released in a sandy area in the vicinity of Waukegan, Illinois. If any survive today they form an isolated colony outside the normal range of the species.

Beyond the Texas border, as we crossed the treeless plains—

originally so devoid of landmarks that pioneers are said to have plowed furrows miles long to provide guiding lines on the flat prairie—we still encountered these wandering, dusty, dry-country turtles. In the years of the great dust storms in the Southwest, those cloudbursts of dry rain merely inconvenienced the creatures. They closed their shells and waited. When the storm was over, they clawed their way upward to the surface through the fallen blanket of dust.

During this summer, with its more than normal rainfall, dust storms were few. We encountered an odd by-product of the year's added moisture when we pulled into a filling station at Dalhart before swinging west on the angling highway that leads to Texline and upper New Mexico. The owner said everybody had been complaining of the heat. Ordinarily the midday temperature at this time of year was about 100 degrees F. This year it was in the low nineties. Yet, instead of feeling better, people felt worse. He attributed this to added humidity in the usually dry air.

The area we now traversed—once Comanche country, mustang country—was part of the Texas Panhandle, rich in oil and cattle. It is the main source of the world's supply of helium, that gaseous by-product of radioactivity that was discovered spectrographically on the sun thirty years before it was first found on earth. As late as 1849, an American army officer reported on a survey of the Panhandle region: "This country is, and must remain, uninhabited forever." That nearsighted observation, however, was no match for the superlative mixture of oratory and ignorance with which the great Daniel Webster opposed the whole development of the West:

"What do we want with this vast and worthless area, this region of savages and wild beasts, of deserts, of shifting sands and whirlwinds, of dust, of cactus and prairie dogs; to what use could we ever hope to put these great deserts, or those endless mountain ranges, impenetrable and covered to their very base with eternal snow? What can we ever hope to do

PARADE OF DUSTY TURTLES

with the western coast, a coast of 3,000 miles, rockbound, cheerless, uninviting and not a harbor in it? Mr. President, I will never vote one cent from the public treasury to place the Pacific Coast one inch nearer Boston than it is now."

As we drove west in the white heat of the early afternoon, all the turtles had taken shelter. Not a man, not a child, not even a dog was abroad on the dusty streets of Texline. Among the dry New Mexican hills beyond, the scattered clumps of the yuccas stood out black-green in the glare of the sunshine. With its sword leaves, lance-tipped, and its spikes of creamy-white, bell-shaped flowers, the yucca is the characteristic plant of the region. Spanish dagger, Spanish bayonet, Eve's thread, Our Lord's candles, Adam's needle, soapweed and soaproot are some of its numerous names. It is a plant of many uses as well as many names.

In pre-Columbian times, the Indians of the Southwest put to use almost every part of the yucca. The cliff dwellers wove their ropes and belts and mats and sandals from the long fibers of the leaves. They ate the fruit, the flowers and the stalk. Boiled or roasted, the fruit pods are said to possess a flavor suggesting that of the sweet potato. The buds of the flowers were boiled to provide a special delicacy. Even the seeds were strung on leaf fibers to form necklaces of beads for the children. And the young shoots of the yucca were considered a medicinal herb. They were mashed and boiled to obtain the juice which was allowed to simmer until it became a syrupy fluid. This was stored away and rubbed on aching joints as a remedy for rheumatism.

It is the long roots of the yucca, which plumb deeply into the earth and enable the plant to survive through relatively rainless years, that provided the American pioneers and the Spaniards before them and the Indians before them with soap. The roots are chopped and mashed and boiled to produce the yucca soap. The Spaniards used it in washing wool. The Indians, long before the arrival of the white men, employed it in their religious rituals. Even today yucca soap is still in demand.

Almost the only sign of life we saw among the yuccas that day was a single jack rabbit that shot off in a great curving course over open ground to our right. At every leap its hind feet kicked up spurts of dust. We saw the curve of its progress marked by thin little clouds of dust that slowly settled to the ground. The long donkey ears that give this western hare its name bobbed up and down as it bounded. In the journals of Meriwether Lewis, leader of the Lewis and Clark Expedition, those ears receive special mention. On the plains of the Dakotas, where he measured the leap of one jack rabbit and found it had covered twenty-one feet, Lewis set down—with all the originality of spelling that the period allowed—the following description of the ears of the animal: "The years are very flexable, the anamall moves them with great ease and quickness and can contach and foald them on his back or delate them at pleasure."

Far away, across the outspread land of Texas, we had glimpsed the blue of mountains to the west. They lifted higher along the horizon as we advanced, hunching in a great barrier across our path. We felt a rising sense of excitement and elation, such as sometimes accompanies the arrival of a storm. We had a feeling of being on trial, of coming to a time of testing. We had begun to climb the eastern edge of the vast western highland that covers a full third of the country.

The road we followed through New Mexico appears on the map like a jagged tear across the far northeastern corner of the state. It led us through a broken land of mountains and mesas across which, in the fall of 1541, Coronado had led his disillusioned followers southward, all their dreams of golden cities destroyed. It carried us through the country of the ancient Folsom Man; through the onetime range of the famous wolf, Lobo, Ernest Thompson Seton's "King of the Currumpaw." It led us, climbing all the way, to the crest of Raton

PARADE OF DUSTY TURTLES 293

Pass on the Colorado border. Through this gap, named by the pioneers for the great number of pack rats along the way, had passed the main flow of the old Santa Fe Trail, going north and going south, in the days of the ox carts.

As we reached the crest, 7,834 feet above sea level, a black roadster pulling a cow pony in a trailer cut around us. The pony looked back as it went by. It stared at us through great, yellow, insectlike eyes—large goggles that had been placed on its head to protect it from the wind. Following this horse with goggles we rode down the long descent and out onto the mile-high Colorado plains.

We found ourselves in a sea of sunflowers. They flooded away in all directions. They washed, to the west, in a kind of floral tide line along the base of the mountain range. Before us now extended all of Colorado, a state that, like Florida and California, holds endless interest for the naturalist. Before us lay the new world of the Central Rockies, the life above timberline, the final weeks of our journey, the high-country end of all our summer wanderings.

TWENTY-NINE

STONE DRAGONFLIES

CLIMBING into the mountains west of Colorado Springs, we came to the village of Florissant and turned onto a dirt road down a narrow valley nearly 8,200 feet above sea level. In the summer sunshine that morning, the valley floor was gay with the red of Indian paintbrush and the blue of gentians. America's most famous mountain, Pike's Peak, towered in massive silhouette fifteen miles to the southeast. Less than twenty miles ahead lay the historic gold fields of Cripple Creek. Here in this pleasant valley, treasure, too, had been mined—treasure of a different sort. In museums around the world Florissant is famous as the valley of the fossil insects. It was the place we wanted to see first in all of Colorado.

Somewhere between 10,000,000 and 25,000,000 years ago, when the waters of a shallow lake covered the land, and palms and sequoias grew along the shore, the sky above was darkened by the volcanic ash of a succession of violent eruptions. Sifting down through the air and water, this fine material hardened into layers of shale, some paper-thin, on the bed of the lake. Between these layers, as on the pages of a giant illustrated book of the past, the insects of the Miocene were preserved as fossils. More than 30,000 specimens, embracing well over 1,000 different species, have been unearthed here. Approximately one out of every twelve kinds of prehistoric insects known to science have come from this one small valley in Colorado.

But the fame of Florissant—pronounced *Flor*-sent by natives of the region—rests not only in the number and variety of its fossils but on the perfection of detail they have retained. Minute features of even the frailest insects, gnats and mosquitoes and soft-bodied plant lice, have been kept through the ages. In some instances, the individual facets of the compound eyes can be detected with a simple hand lens. In others, the feathery gill tails with which nymphs of some of the smaller dragonflies obtained oxygen from water of the Miocene have been preserved in the minutest detail. Dragonflies of stone, fossil insects that darted about on veined wings in the sunshine of at least 100,000 centuries ago, have been found in half a dozen species in the layered shales of Florissant.

Fossil trees rather than fossil insects originally attracted attention to the valley. From the Ute Indians, early pioneers heard tales of great white trees of stone. They investigated and found the petrified stumps and trunks of immense sequoias, one of them credited with being the largest fossil tree in the world. In the early 1870's the first scientific group, a party of government geologists, explored the valley. Soon afterwards, Dr. Samuel Hubbard Scudder, entomologist of Cambridge, Massachusetts, and the American authority on grasshoppers, began a serious study of the insects of the Florissant shale. For decades he delved among these creatures of the Miocene. In all, Scudder gave names to 233 genera and 1,144 species of fossil insects. His classic monograph, *The Tertiary Insects of North America*, appeared in 1890.

In the intervening years, many men have come to this insect Pompeii to search for victims of those ancient volcanoes. Their finds are scattered through museums and centers of learning in many parts of the world. At present the "white trees" of the Utes, the "petrified stumps" of the pioneers, the fossil-bearing shale of Scudder, can be visited in two adjacent areas, the Colorado Petrified Forest and the Pike Petrified

Forest. They lie about thirty-five miles west of Colorado Springs, near the southern extremity of the Front Range of the Rockies. Both are privately owned. For a relatively small fee, visitors can examine the trees and stumps of stone, can dig among the layers of shale and can even keep whatever specimens they find. The deposits of the ancient lake underlie many square miles of the valley, but in these two places the fossil-bearing layers are most readily accessible.

We came to them first that morning, and we were drawn back more than once during our days in Colorado. We walked among the petrified trees. We inspected the largest fossil of its kind on earth, a sequoia stump seventy-four feet in circumference. But it was the layers of shale, ranging from white to chocolate-brown, that absorbed our attention most.

Where embankments had been cut away, the strata lay exposed like the pages of a book lying on its side—pages of unequal thickness. In separating extracted pieces of the layers, visitors have employed a wide variety of impromptu aids—screwdrivers, pincers, nail files, shovels, chisels, jackknives and razor blades. On our first visit, we used a pocketknife; later we brought along a sharp-pointed prospector's pick to help pull out portions of the shale. The rock is rather soft and the most productive layers split apart easily. Not infrequently, the strata were so weakly cemented together we could pull them apart with our hands. When the shale is dry, we soon discovered, the fossils are easier to see.

The richness of the Florissant deposits has been a constant source of amazement to scientists. During one summer in the 1870's, Scudder obtained more than twice the number of fossil insects that the German scientist, Heer, found in thirty years of searching at the famed Bavarian quarries of Oeningen. In 1912, Professor H. F. Wickham, of the University of Iowa, dug a trench about twenty feet long and six feet deep. From it he obtained well over ninety species of beetles, more than forty of them new to science. On several occasions, pieces of shale no larger than an outspread hand have contained several insects preserved close together. This does not mean, of course, that every piece of shale—or every thousand pieces of shale—split apart will contribute a prize.

But nobody knows what will come next. *This* may be the one! Digging for these mementos of the past is like picking up sea shells on a strange shore. It becomes an engrossing game. We lost track of time. We dug with a kind of fossil fever, prospector's excitement. With a pocket magnifying glass, we examined each spot and stain. Scudder found that the most numerous of all insects at Florissant—as they are in the world today—were the ants. He collected more than 4,000 ants of fifty different species. They represented about 25 per cent of his total. A dozen times, we thought we had discovered ants. But the magnifying glass refused to be fooled. It expanded the small, dark spot into a bit of leaf or bark.

For fossils of the prehistoric vegetation are the most common of all. In various collections, I have seen the leaves of ferns, of roses, of iris, of elm and chestnut and poplar, sumac and tree of heaven, of balloon vine and the Oregon grape, all obtained from Florissant shale. A complete cattail head, on which some ancient dragonfly may have rested, is imprinted on one rock. Nuts and pine cones and rosebuds and even the delicate petals of wildflowers are numbered among the Florissant fossils. Living representatives of some of the plants now grow in China, Mexico, Norway and the West Indies. Between the layers of shale have been found the skeletons of fish and birds—one a finch, another a plover—as well as fossil tracks and fossil feathers and the small petrified shells of fresh-water mollusks.

Each time we split away a layer of shale and revealed some fragment of an ancient plant, we were letting the first ray of light strike it in millions of years. It had been entombed, just where we found it, since before the ice ages. On the afternoon of our first day at Florissant, a piece of light-hued shale produced our best botanical find. It split apart in my hands and revealed a perfect leaf, elongated and serrated. It seemed, at first glance, to have come from a birch. Later, when I compared it with the collection at the University of Colorado, at Boulder, I found it had grown on a prehistoric water elm.

As we looked up from examining this leaf, we were surprised to see a buffalo with a tan-colored calf leading twenty or so black Angus cattle across a neighboring pasture strewn with blue gentians. I glanced from the gentians to the bison to the leaf contained in the rock I had just picked up. Time was linked together in a great triangle by the flower of today, the bison, symbolic of another generation, and the leaf that had formed its chlorophyll under the sun about which this globe had whirled millions and millions of times before the fossil of its tissues had come to light again. That fossil is preserved—as are all the specimens at Florissant, both insect and plant—in the form of a thin film of carbonaceous material.

It was not always that such objects aroused scientific interest alone. During the long Dark Ages of the human mind, fossils excited fear and superstitious dread. They were believed to be the product of some mysterious "plastic force" or, more often, "devices of the Devil." The eighteenth century was well along before the true nature of fossils was widely recognized.

Among the early collectors of these relics of the past, there appears the somewhat pathetic figure of the German schoolmaster Johannes Beringer, of Würzburg. An avid fossil hunter, he took his students on frequent field trips to a nearby outcropping of soft shale. One day, as a joke, his pupils carved the image of an animal on a bit of rock. Their credulous teacher pounced on it with such enthusiasm that the hoax continued. Each subsequent field trip yielded greater treasures —frogs, flowers, insects, animals, even astronomical objects. In 1726, Beringer published his *Lithographia Würceburgensis*, an elaborately illustrated volume describing all his finds.

Shortly afterwards, he discovered rocks containing Hebrew letters and then one with his own name inscribed on it. This was more than even his credulity could stand. Too late he realized he had been duped. In an effort to buy back every copy of his book, Beringer spent his life savings and died in poverty. Even so, his story has the ironic semblance of a happy ending. After his death, his family recouped their fortunes by selling the copies of the book as a rare curiosity.

Each time we dug at Florissant, grasshoppers danced in the sunshine before the walls of layered shale. They jiggled up and down as though dangling on rubber bands. Red-shafted flickers called and flew over us, turning in the air and bursting into flame as the sun struck the under plumage of their wings. Pigmy nuthatches, red crossbills and Steller's jays alighted in the scattered lodgepole pines—here in the open more widespread and bushy than elsewhere and known locally as "jack pines." The air above us was filled, from time to time, with the sweet music of mountain bluebirds calling and once, coming from the west, a golden eagle soared by, huge on outspread wings, a fit prototype for the Thunderbird of the Utes.

When Samuel Hubbard Scudder was bestowing scientific names on his Florissant fossils he christened one in honor of "the industrious entomologist of Colorado," T. D. A. Cockerell. Of all those who have explored among the shales of this mountain valley, without doubt the most remarkable was Theodore Dru Alison Cockerell. Born at Norwood, a suburb of London, on August 22, 1866, he was interested in natural history from childhood. His first scientific experiment, he recalled many years later, was to disprove the theory confided to him by a grown-up that yellow primroses planted upside down would come up pink. When his father died in his early thirties, young Cockerell went to work for a firm of flour merchants. Never robust, he contracted tuberculosis in the dusty atmosphere and, in the summer of 1887, at the age of twenty, he sailed for America and the mountain climate of Colorado. It was this misfortune early in his long life—he died at the age of eighty-one—that led him into the natural history of another continent and ultimately into the prehistoric world of Florissant.

Without specific training, without any academic degree until he was honored with a doctorate of science long after he had assumed a teaching position at the University of Colorado, this friend of Alfred Russel Wallace wandered through the fields of natural science with the glorious, unspecialized freedom of the old-time naturalists. He became recognized as an authority on wild bees, scale insects, sunflowers, mollusks and fish scales as well as on insect paleontology. Once, while attending a banquet, Cockerell discovered a new species of insect on the bouquet of flowers set on the table before him. During his lifetime he contributed more than 3,500 papers to publications in many parts of the world. Once he achieved the goal of producing a scientific paper during each of the 365 days of the year. He has been well described as one of the most versatile biologists America has known. Bronze plaques commemorating his work have been hung at the museum of the University of Colorado and at the British Museum in London.

In 1890, when Scudder ended his Florissant studies, it was generally assumed that the Colorado shale beds had been worked out. Thus it was not until the summer of 1906 that Cockerell first visited them. He soon decided the beds were almost inexhaustible. Year after year, he and his wife, Wilmatte Porter Cockerell, spent part of each summer exploring among insects of the remote past. During many of these expeditions, they found an average of a new species a day.

They unearthed a tsetse fly, long extinct on this continent but in Africa still the dread carrier of sleeping sickness. They found the leaves of several wild roses and once, when they split open a layer of shale, they came upon a prehistoric rosebud. Their great hope, for years, was to discover a fossil butterfly. The only specimens known to the Western Hemisphere have come from Florissant. After years of searching they were returning to the village one day by a new path. They stopped to rest where a bit of shale protruded from a hillside. Mrs. Cockerell turned over a piece and there was the long-sought fossil, a butterfly so perfectly preserved that the spots were still apparent on the wings. Cockerell named it, in honor of its discoverer, *Chlorippe wilmattae*.

In further studies of prehistoric insects, Cockerell and his wife traveled to Argentina, to Europe and to Siberia. Late in life, they journeyed up the whole length of Africa from Cape Town to Cairo. The interest they found in nature, and in the scientific exploration of it, expanded before them wherever they went. "The scientific man," Cockerell once wrote, "is always on the road, never at the journey's end."

The great collection of Florissant fossils that these two made is now housed in the Museum of the University of Colorado. When, later on, we visited this institution at Boulder, we saw spiders and earwigs and lantern flies, cicadas and back-swimmers, a wasp that resembled a yellow jacket and a weevil with a slender snout, not unlike the beetles we find in roses today. A number of the prehistoric leaves contained the swellings of insect galls, and a small sheet of rock held the larval case within which a prehistoric caddis fly had lived in the shallow lake at Florissant. We were shown one fossil leaf from which little circular pieces had been cut, snipped out at least 10,000,000 years before by the jaws of a leaf-cutter bee. The preponderance of spring forms among the fossil insects contained in one layer of Florissant shale reveals the very season of the year when one of the volcanic eruptions occurred. All told, the Cockerell collection contains more than 200 type specimens, the original finds from which new species were described.

In our own search for a fossil insect, we continued to be disappointed. I discovered the fragment of a root. Nellie found some kind of a small seed. We both uncovered bits of leaves and portions of twigs. Each time we sank the prospector's pick into the soft rock and pulled out a plate of shale to split apart and examine, we hoped to expose some insect of the Miocene. We tried hard to make insect wings from bits of leaf. But we were never completely convinced.

Then I found it! I inserted the thin blade of my jackknife carefully between two layers of gray shale. They fell apart. There, revealed where it had lain since before there were human eyes on earth to see it, was the clear form of a large crane fly. One of its wings was almost perfectly preserved. The other was partially crumpled and damaged. In the long-ago days when it fluttered over the lake shore among the sequoias and palms, the span of those wings must have been at least an inch and a half. We examined our prize over and over. There was the body and there the head and there the widespread, sprawling, stiltlike legs characteristic of all the crane flies.

We sat in great content under one of the pine trees that afternoon. We let our eyes wander over the pleasant scene around us, over the thin spires of the green gentians, over the tansy asters—still blue, still pungently aromatic when crushed —over the rose-pink cranesbill with its geranium-like bloom, over all the quiet beauty of these latter days of summer. Relaxed, we listened to the little sounds around us, to the flutter of the dancing grasshoppers and the low croon of the breeze among the pine needles. Immense cumulus clouds, snowy white in the sunshine, piled continually higher in the sky.

Just so, other clouds, black and tumbling and volcanoborn, had once billowed up here. They had darkened the sky, bearing their burden of ash. We tried, as we talked, to reach back across the millions of years and visualize that day of destruction, the last day in the life of the crane fly. Then everything we now saw had been reversed. The sunshine had been extinguished. The day had been turned into night. The tumbling clouds were black instead of white. The clear air was choked with the endless, sifting fall of fine volcanic ash. Under its hardening mantle, the crane fly had slowly altered into fossil form. There it had remained unchanged through millenniums and eras, ages and epochs, while mountain ranges rose and coast lines altered and ice ages came and went. And now, at last, on this late-summer day of sunshine, we had split open two sheets of shale and the crane fly within had returned once more to the world of light.

THIRTY

HIGH TUNDRA

MOUNTAIN flower fields. Timberline gardens. The high tundra in bloom. One of the most dramatic events of summer is the flowering of the mountain heights when Alpine plants bloom all across the lofty meadows above timberline. This we had never seen. This we had long hoped to see. And now it was all around us.

We had ascended the Trail Ridge Road that climbs to a height of 12,183 feet in crossing the Rocky Mountain National Park in the north-central part of Colorado. For four miles it runs above 12,000 feet, and nearly one-fourth of its forty-eight miles is above timberline. Following roughly an ancient trail of the Ute Indians, it is the highest continuous automobile road in America. During the summer, when the mountain flower show reaches its height in July and August, the Trail Ridge Road is the country's most spectacular path into the world of Alpine blooms. It leads across eleven miles of tundra in the sky.

For above timberline the sweep of wiry grass and stunted herbs is comparable to the tundra beyond the tree line in the far north. The climate of these mountain tops is similar to that of Siberia and upper Alaska. We had climbed into such surroundings as we might have found beyond the Arctic Circle, at a latitude of seventy degrees instead of the forty we occupied. During our ascent from Estes Park, the temperature had dropped about three degrees for every 1,000 feet we climbed, a total of nearly fourteen degrees. Gone was the heavy heat of the Great Plains. The sparkling, rarefied air was pleasantly cool, in spite of the blaze of the sun, as we wandered over the high meadows amid the unfamiliar flora of the Arctic plants.

August was departing. Blooming time was drawing to a close. We were a little late. But, among the mountains, this particular summer season was reported to be nearly two weeks behind normal. This was our good fortune. For, above timberline, plants were still rushing into bloom, swiftly maturing their seeds. Even though we had arrived near the end of the show, the glory of the Alpine meadows, of mountains in flower, spread about us.

Henry Thoreau concluded that, in the vicinity of Walden Pond, in eastern Massachusetts, nine-tenths of the wildflowers had finished their blooming by the end of July. Here, some 1,700 miles to the west, where the high-country summer was drawing so close to its early autumn, we were seeing the last small fraction of the blooms of the year. Yet how rare, how strange, how beautiful they were!

Each time we visited these treeless heights, under the drifting clouds, surrounded by mountains shining in the sunlight and mountains dark in the shadows, we made new discoveries among the wildflowers. Tufts of white, some turning pink as they grew older, were scattered widely on slender stems. They were the flowers of the bistort, a buckwheat of the high mountains. Yellow five-petaled blooms rose above leaves deeply toothed and purplish-green. They were the Arctic avens. All the cream-colored flowers of that Alpine member of the snapdragon family, *Chionophila jamesi*, the snow-lover, were concentrated—like half the feather on an arrow—along one side of the supporting stem.

Frequently we came upon the clustered floral goblets of the Arctic gentians. In a single clump, we found eighteen massed together, each greenish-white, streaked or spotted with purple. What appeared to be escaped red clover growing in dwarf form on the mountain top caught our attention. It was the miniature rose or Parry's clover of the high meadows. Beside one Lilliputian pool on the tundra, a clump of plants lifted on short, silvery stems masses that seemed formed of dark red beads. They proved to be the almost opened buds of a sedum or stonecrop, the king's crown, *Sedum integrifolium*. A little farther down the same slope, we came upon another sedum, the rose crown, pink instead of richly red. And each day, as the sun sank lower and we looked toward it down a slope, we saw the bracts of the yellow paintbrush turned to golden-flamed torches in the backlighting. The red paintbrush predominates at lower levels, the yellow above timberline.

According to Dr. P. A. Rydberg, a specialist in the group, there are approximately 250 species of plants belonging to the strictly Alpine flora of the Rockies. In addition there are about 100 other species that occur above as well as below timberline. Few plants survive with more tenacity than these high-mountain species. Almost all are perennials. Summer is too short to complete their life cycle in a single season. In the case of some Arctic plants, years go by during their slow advance from germination to a first blooming. We gazed in awe at these small, frail-appearing, enduring plants, their slow progress gained, their growth so hardly won, now at last in the blooming time and seedtime of their lives.

Most of them hugged the earth, miniature plants with miniature blooms. We bent to look closely at a tiny spray of yellow goldenrod. It was dwarfed to a height of no more than three inches. In the shelter of a stone, Nellie discovered a minute willow—I believe it was that smallest of its kind, the Rocky Mountain snow willow. Its top rose hardly more than the length of my forefinger above its roots. Yet it was covered with white and fluffy catkins, some of them dangling down for almost half the height of the tree.

Not infrequently, the Arctic plants huddled together in dense clusters or mats. Strewn across the cushions of the moss campion were pink flowers, across those of the Alpine sandwort white flowers. Where one mat of moss campion had already bloomed and gone to seed, the low, dense plant mass, eighteen or more inches in length, appeared to flow like viscid fluid over and down between two small rocks pink with feldspar. Experiments have shown that mats of moss and darker plants, high on a mountain top, absorb sufficient solar radiation to produce at times a microclimate of their own. In one instance, the temperature of the air a few inches from the ground was only ten degrees above zero, while within a mat of dark-hued plants it was fifty degrees F.—a difference of forty degrees. Thus some of these densely clustered plants are able to grow even when the temperature of the air around them is below the freezing point.

At every step we sank into the springy layer of turf beneath our feet. How many ages of slow development from the humus of moss and lichen, from the deliberate, infinitesimal erosion of rocks, had produced the first of this mountain-top soil! Great rocks, pink-tinted, thrust up through this thin skin of vegetation. And always the cloud shadows swept across us, across the great rocks and the little rocks scattered everywhere around us, across the flowers of the tundra sod.

The Alpine bloom we wanted most to see was the little red elephant flower. Its history is curious. It was originally discovered on Greenland as its scientific name, *Pedicularis groenlandica*, indicates. Later it was found growing under the Arctic conditions of high mountain tops in the Rockies. The first of these fairyland flowers we came upon was rooted between twining threads of seeping water that advanced almost imperceptibly down the incline. Above its fine-cut, comblike leaves, it lifted a purplish-pink spike formed of many flowers. We bent close and examined it with mounting delight. Each little flower resembled a miniature elephant's head complete with flaring ears, bulging forehead and down-curving trunk. What that long-ago botanist had first seen on Greenland's Arctic tundra, we were seeing on this mountain summit in mid-America. Although separated by thousands of miles, the little red elephant flowers of Greenland and those in Colorado bloom at the same time.

Many years ago, in an Alpine garden in Tyrol, the great student of living plants, Anton von Kerner, conducted exhaustive studies of the changes that take place when lowland plants are transferred to higher elevations. He found that those that survived tended to shorten their stems. The color of the petals became more intense. And most surprising of all, in spite of the fact that the growing season begins much later in the high meadows, the same species may bloom earlier on the heights than in the lowland. In the case of such European plants as *Gentiana germanica* and one of the *Parnassia*, the flowers opened among the foothills in August while, in the mountain meadows of the upper Alps, the blooms appeared in July. Mountain tops receive direct sunlight, through thin atmosphere, with few of the rays filtered out. There the life of the plant is speeded up.

One spot on the tundra we visited again and again. We left our car in a tiny mountain turnout and climbed upward for half a mile or so, first among the contorted trunks of timberline, then out upon the tundra and finally beside a lonely little snowfield nestled at the foot of a rock slide. Here we were at the edge of a great saddle, 12,000 feet high, alone, completely, intensely alone, with the rugged grandeur of the mountains spreading away peak on peak around us. We never saw signs of another human being. We never saw a footprint here other than our own. Always we had the snowfield and the flower-carpeted tundra to ourselves. It was our own special, magic place.

Here the Alpine flowers bloomed in greatest variety. They pressed close to the snowfield; they streamed away down the saddle slopes; they climbed above us, circling the fallen talus and ascending the tundra to higher, greater rocks where the marmots lived. Little flowers, demure flowers, bright flowers, flowers that were all wonderfully new, led us on and on over the treeless heights.

We often sat on low stones to see, closer at hand, the small Alpine plants around us. In two square feet of the mountain turf, I once found the yellow paintbrush, the rose-purple of little clover, the white of bistort and the yellow of the dwarf senicio. Close to timberline, Nellie discovered the pale blue flowers of Jacob's ladder, a plant that first blooms on Long Island in the spring but here was in a late flowering on the dividing line of September. Only once did we see a wand lily and rarely did we catch sight of the yellow avens.

Only a few weeks before, the Alpine avens had been the dominant flower of the high meadows. During our successive visits we found, when several days intervened, marked changes in the blooming plants. One time we returned to our little snowfield and discovered the white of new bistort strewn thickly over wide patches of the tundra. Again, green gentians seemed to have sprung up like mushrooms all down the damper stretches of the slope.

The snowfield itself changed and shrank from day to day. It was not part of the perennial snow of the high mountains. Rather it was a melting fragment of a winter drift. Close beside it we beheld the creamy yellow blooms of our first Alpine anemone. And above it, at the time of our earliest visit, a great clump of Colorado columbine, the flower of the state, leaned out over the rock slide, its pale-blue and white flowers, each fully two inches across, nodding in the breeze. We counted seventeen of these delicate, gorgeous flowers in this single cluster.

And as we gazed our delight was suddenly multiplied. For a large sphinx moth, marked with pink and subtle shadings of gray, materialized in front of the flowers. Before each in turn it hung hummingbird-wise, caught as by a spotlight in the full brilliance of the mountain sunshine. Bees and bee-fertilized flowers are largely left behind on the heights. Blooms pollinated by other insects increase.

Two miles and more above sea level, the insects we saw appeared darker in hue than species lower down. One large dragonfly, almost black, went sailing by. Even the butterflies seemed more dusky of wing. This characteristic has been noted both among Arctic and Alpine insects. Their darker colors absorb more of the heat of the sun.

In seventeenth-century London, Samuel Pepys set down in his now-famous diary the following entry: "23rd. In my black silk suit (the first day I put it on this year) to my Lord Mayor's by coach, with a great deal of honourable company, and great entertainment. At table I had very good discourse with Mr. Ashmole wherein he did assure me that many insects do often fall from the sky, ready formed."

And so they do—not because they are generated in the atmosphere as men of Pepys' day believed, but in accordance with the simple rule that what goes up must come down. Swept aloft by powerful updrafts, even wingless species are sometimes carried far. During summer months there is a continual floating population in the air and a constant rain of life from the sky. Especially is this true among the mountains.

For twenty years, Dr. Gordon Alexander of the University of Colorado studied the altitudinal distribution of grasshoppers in his state. He found that these insects outnumber all others on the high meadows of the Alpine Zone. Yet of the twenty-eight species and sub-species collected there, only eleven can be classed as residents. All the others have been carried up the slopes and deposited by the rising winds.

One dramatic consequence of these grasshopper winds of the Rockies is the formation of grasshopper glaciers. When the insects are deposited on permanent fields of snow and ice, they become numbed, frozen and often embedded in the glacier. The most famous site of the kind, in the Beartooth Range of southern Montana, a wall of ice eighty feet high and a mile long, is streaked with successive layers of insects. The lowest stratum, buried under sixty feet of ice, is centuries old. During the hottest days of summer, melting releases some of these insects from cold storage, and rosy finches—and even bears—come to feed on a banquet of grasshoppers.

Rosy finches alighted, too, on our little snowfield in search of insects. They were the brown-capped rosy finches, beaver-brown with varying shadings of red. Almost their entire breeding range is confined to Colorado. They nest above timberline and winter in the valleys. Several times, little family groups drank, within thirty feet of us, at a thin rivulet of melt-water that ran off down the slope. Several of these birds were not long from the nest. They opened their bills and fluttered their wings and followed other finches begging for food.

In several places on the melting snow we noticed small dead insects with little pits beneath them. Their darker bodies had absorbed more of the heat of the sun and concentrated melting there. It was similar pits—deeper and larger, formed beneath dead grasshoppers—that enabled the first party of Government geologists to reach the top of the Grand Tetons, in Wyoming. By using these insect-produced pits as finger holds, they pulled themselves up the last steeply-tilted sheet of ice to attain the summit.

Once, when we were bending over an Alpine flower, we heard a sudden chiming of sweet bird warbles—all the finches calling. We looked up. On slender, pointed wings, pale sandy-hued with dark patches under the base of each, a prairie falcon skimmed up and over the saddle, saw us, swerved away down above the treetops of the timberline. It was another hawk, I think, that surprised us most. Tilting this way and that, white rump patch appearing and disappearing as it turned, a marsh hawk came beating low above the tundra. Each year among the Rockies, this bird—associated in our minds with lowland fields—makes an altitudinal food migration, ascending higher and higher between August 1 and September 15 until it is foraging for mice along the summits of the mountains.

We speak of mountain weather. We should, no doubt, speak of mountain "weathers." For where else are the varieties so great, the changes so sudden? One afternoon, as we wandered above timberline, the sky darkened swiftly. A chill wind raced ahead of the storm. Thunder, in the wake of broad flashes of lightning, rumbled on and on, caught among the peaks. Beneath the black clouds, long, descending veils of rain or snow drifted in ever-parting curtains that, in turn, obscured and revealed the distant slopes. Rain was falling when we left the tundra; snow was driving around us when we reached the wind timber of tree line; hail was descending when we found a sheltered nook and a seat on an ancient spruce trunk.

Sitting there, we listened to the changing music of the wind among the branches and needles of the Engelmann spruce. Among trees the force of the wind swiftly abates. Measurements have shown that 1,000 feet within a forest the velocity of the wind drops to one-tenth its speed in the open. Around us, sheltered by the twisted trees, the delicate blue of chiming bells ran in clusters and waves along the slope. We had been there hardly a quarter of an hour when the clouds and the trailing veils of snow and rain moved away toward the east. The sky was swept clear. The sun shone. A yellow, black-capped piolated warbler darted among the spruce needles. All in less than an hour's time, we had experienced sunshine, clouds, veils of falling vapor, rain, snow, hail, lightning and thunder; black, swollen sky, clear air and sunshine again.

At the close of another afternoon, toward sunset, we sat on a lichen-covered rock above the clump of Colorado columbine and ate our supper of sandwiches and large sweet blue plums. Over us, American pipits flew, calling their names, to alight on boulders or walk over the ground, brownish birds wagging their white-edged tails. Behind us, low shafts of sunshine streamed out of a broken sky. They picked out, as in a spotlight, distant peaks and slopes. From where we sat, we could look down over the saddle on the timberline of the mountains ranged about us. We could see the forests climbing the steep ascents, slowing down, bending forward, falling prostrate and ending in a ragged line of matted needles and contorted branches pressed against the ground.

In this area, timberline occurs at about 11,500 feet. It varies 1,000 feet between northern and southern Colorado. Where strong winds blow, the timberline may be depressed as much as 1,500 feet. Looking away below us, we could see the tree line varying with local conditions from mountain to mountain. It stood lower on the shaded than on the predominantly sunny slopes. It rarely ran in a straight line. Tiny advantages and disadvantages are sufficient to tip the scales of life and death in the harsh struggle that is unending along these upper limits of the trees.

After a time, one of the shafts of light swept the tundra slope behind us. The flowers turned luminous. Among them all, across the turf, among the rocks, in the golden mist of backlighting, shining yellow disks like sunflowers stood out. They were the single blooms of the Alpine goldflower, the sun god, the old man of the mountain, *Rydbergia grandiflora*. Some of these flower disks were four inches across, larger than many of the sunflowers on the dry plains 7,000 feet below. Furry gray stems lifted the blooms only a few inches above the ground. Anton von Kerner, long ago, pointed out that high-mountain plants, exposed to the wind, often have woolly stems and silky or felted foliage. An outer layer of sapless, air-filled, interwoven hair structures provides protection against too rapid evaporation.

As we roamed along the slope, photographing the glowing flowers, we came upon an elk horn gnawed by mice in search of calcium. At times we found ourselves in a maze of wandering, interconnecting cylinders of solid sand and pebbles. In cleaning out and enlarging their burrows at the end of winter, pocket gophers had packed the excavated material into tunnels they had made under the snow. When summer comes and the drifts melt, the resulting labyrinth of earth cores—two or two and a half inches in diameter and sometimes covering hundreds of square feet—stand out in pale yellow on the green of the tundra.

Among the jumble of great rocks that rose a quarter of a mile above us, marmots called back and forth in the sunset. In that thin, far-carrying double-note, we heard some of their last whistles of the year. It was now only a matter perhaps of days, of two or three weeks at most, before they would fall into the profound, deathlike winter sleep that occupies seven of the twelve months of their year.

Steadily the shadow of the mountain behind us crept across the tundra saddle. Around us our last sunset on the heights slowly faded. By 6:30 P.M., the wind had fallen, the birds were gone, the marmots were silent. The stillness of the summit became so intense our ears seemed to hum. The soundless mountain top resembled a country village shut up tight for the night. We started down, winding like mountain goats among the rocks, crossing a marmot run and an elk trail on our way to timberline.

When first we came to the unfamiliar world of these high peaks, they left us feeling small and lost, overawed and alien. Then we made friends with the little things, the bistort and the chiming bells, the rosy finches and the pipits, the moss campion and the snow willow, the avens and the deep purple bells of the penstemon clusters. We became familiar with fragments of the whole. We went, bit by bit, from the particular to the general. And thus, advancing inductively from the small to the large, we came to know and love the mountains.

THIRTY-ONE

SUMMER SNOW

IN THE spring, on Long Island, we had selected this early-September day. We had written our friends, Dwight and Esther Pennington, that on the morning of this day we would meet them at Grasshopper Park, near Deadman's Gulch, high in the Gunnison National Forest on the western slope of the Great Divide. The morning had come and we were bumping along the rocky, single-track road winding beside Spring Creek, climbing to 9,000 feet above sea level.

We were 100 miles west of Florissant and its fossils, 130 miles southwest of the Trail Ridge Road and its high tundra, almost in the exact center of the Colorado Rockies. The day before we had descended into the wide, flat expanse of South Park, the lowest of a chain of four great basins extending down the center of the state behind the Front Range. We had climbed to 11,312 feet to cross the Continental Divide at Monarch Pass. We had come to the crest in rain and had made the long run down the western side in clearing weather and had ridden into Gunnison in the early dusk with nighthawks hunting moths low above the street lamps.

Now over a wooden bridge, at a turn in the mountain road, we came to the Penningtons' summer cabin. It was made of logs and set in a wildly beautiful spot. More than thirty years before, when Nellie and I had gone to Wichita, Dwight and Esther had been in college. They had been our students then and our friends ever since. During the intervening years, while

Dwight was following a straight path to the editorial page of the *Kansas City Star* and I had been wandering widely, our trails had repeatedly crossed and diverged and met again. Sitting now beside the high-country stream beneath an immense Douglas fir, we talked over other days and other friends. And as we talked, we refreshed ourselves with honey-sweet, brook-chilled Rocky Ford muskmelons.

The Penningtons had built their rustic cabin more than twenty years before. While it was under construction, pack rats stole their toothbrushes and piled nesting material, night after night, on top of the battery case of their Ford. And a porcupine gnawed a hole in a brand-new tire. More recently beavers had offered violent resistance to their scheme for obtaining mountain water under pressure. Three hundred feet of polyethylene tubing was run up the hillside along a little branch brook to the first of a series of beaver ponds. The lower end of the tubing was attached to a faucet in the cabin. The upper end, with a strainer-funnel on it, was placed in the pond. An ample supply of running water, under pressure, resulted. That is, it resulted for a day or so. Then the beavers discovered that in some mysterious way, although their dam was intact, their pond was losing water. They promptly gnawed the line in two.

Upstream from the cabin, Dwight led us to a dead spruce that leaned out over Spring Creek. Long since it had lost its bark. The grain of the bare trunk, instead of ascending normally, rose in a continual spiral. The fibers followed a corkscrew path so that the tree suggested a rope of twisted taffy. Once more we were seeing an arboreal riddle that had interested us in many places along the way.

Close beside Devil's Tower, in Wyoming, one lofty dead tree had every barkless branch from top to bottom, as well as the central trunk, formed of tightly spiraled wood. Somewhere in the Rockies we passed a cabin that was made entirely of spiral logs. Botanists have found that the twist of the

ENGELMANN SPRUCE near timberline. On the following two pages, the Gunnison country near Emerald Lake.

TIMBERLINE TREES, above. Below left, a storm-wracked spruce. At right, a dead tree exhibits spiral grain.

grain may be either to the right or to the left, although the right-handed turn seems more common. The angle of the spiral, and even in extreme cases its direction, may alter with age as the tree continues to grow. When lightning strikes one of these twisted trees during its lifetime, the thunderbolt, instead of grooving out a vertical downward path, descends the trunk on a corkscrew course.

What is the explanation for this spiral growth? What causes such trees? Precise answers to those questions are still to be found. Many suggestions have been made; many hypotheses offered. One theory is that windstorms, when the trees are young, start the cells growing in an abnormal way. I remember once being assured that such trees result when the branches on one side are longer than on the other side, thus giving the prevailing wind leverage to turn the tree slowly as it grows. How transparently unsound this explanation is can be seen if you follow one of the longer branches through half a turn. Then it would be on the opposite side of the trunk and would be moving against the force of the prevailing wind instead of with it.

Usually trees with spiral grain are found in harsh surroundings. We saw them most frequently near timberline, where the fight for survival is unending. However, the fact that one tree will be spiraled and another close by will not shows that the explanation lies within the tree itself, that it is produced by some idiosyncrasy, inherited or individual. Recently, radioactive isotopes introduced into growing trees have revealed that sap follows a more spiral course than had hitherto been suspected. "Apparently," as one botanist sums up the present state of our knowledge, "the spiraling tendency in trees extends to the fibers under certain circumstances as yet not clearly understood." Thus the mystery of the bird's-eye maple is matched by the riddle of the twisted trees.

In all this Gunnison country, as we roamed over it during the succeeding days, we were in a land of picturesque names:

Tincup Pass, Tepee Gulch, Needle Rock, Soap Basin, Mysterious Lake, Flatiron Mesa, Porcupine Cone, Terror Creek and Holy Terror Reservoir. Once we passed a Whetstone Mountain and another time an Oh-Be-Joyful Creek.

Dwight's Jeep, a nine-year-old veteran of the mountain roads, carried us, on one side trip, up over rocks and ruts of a primitive road, past marmots whistling from great boulders, under trees gray with beards of usnea lichen, to a small, two-man manganese mine, 10,000 feet high on a mountainside. Close to the top of our climb, a robin flew past us. All along the Rockies—those titanic wrinkles on the face of the continent produced by east-west pressures—birds were on the move these days. Migration begins early in the high country. By mid-August the nighthawks are drifting southward, hunting insects as they go, and the broad-tailed hummingbirds are working from mountain meadow to mountain meadow down the long chain of the timberline gardens.

Just as we had descended from Wilkerson Pass into the valley of South Park, on our way to Gunnison, we had seen one of the last of these nectar feeders following the high road of the hummingbirds. It had paused momentarily, in a glitter of metallic greens, at a clump of wildflowers beside the road. Then it had streaked away into the south. At a speed of fifty miles an hour, with its wings beating seventy-five times a second, it was on its way toward its far-off wintering ground in southern Mexico.

Many times as we walked under the stars on September nights, we heard the small cheeps or calls of migrants coming down from the sky. Some of these late-summer travelers among the Rockies first fly east and then turn south along the edge of the Great Plains. Others move directly south among the mountains. Still others, such as the gray-headed juncos, the brown creepers and the white-crowned sparrows, make a vertical or altitudinal migration like the Carolina juncos we had encountered among the Great Smokies in the spring. When colder weather comes, they migrate downward to winter among the foothills and bordering plains of eastern Colorado.

Cliff swallows and red-winged blackbirds and crows were flocking all down a long valley when Dwight and Esther took us, next day, on a 180-mile loop through some of the most wildly beautiful and rugged scenery on the continent. It carried us as far west as the Black Canyon of the Gunnison.

Before we came to Swampy Pass or Ohio Pass or Kebler Pass, we saw on our right a tan-colored mountain with slight tintings of rose in the sunshine. It was Carbon Peak. Steeply it lifted to the naked rock of its upper slopes. There the strata tilted sharply downward. Over a period of years, gravity dragging on layers of shale that rested on layers of sandstone had shifted their position noticeably and had made the peak famous as Colorado's "walking mountain." Another oddity of movement found among the Rockies, especially in the steep valleys of the San Juan Mountains of southwestern Colorado, is the rock glacier. Where immense masses of broken fragments collect on an incline at the base of a peak, they are sometimes impelled by gravity and the freezing and thawing of water within them to make a slow, gradual advance year after year, creeping down the slopes of the valleys like the ice mass of a glacier.

Summer hay was now in stacks in the fields of the mountain ranches, each stack surrounded by a temporary fence to keep the cattle—and deer—away. One by one these fences would come down as the last of the range grass disappeared. Summer's end was approaching. We saw the fact indicated in many places along our way. We saw spiders among the rocks carrying egg cases and were reminded of all the seeds of life being planted by small creatures during these latter days of the season. A little below 10,000 feet, descending the far side of Kebler Pass, we came to a mountain meadow filled with sheep. They fed for half a mile along the grassy slope. A sheepherder, with a string of burros, was moving the flock, day by day, down from the highest pastures. This annual migration of the sheep is as sure a sign of the approach of autumn as is the flocking of the birds.

When we stopped for a magnificent view where the seamed eminence of Mt. Marcellina ascended beyond miles of sunflowers gone to seed, I walked out among the dry plants. All around me there was the scurrying and squeaking of ground squirrels. Every rock was covered with a carpet of hulls and fragments of sunflower heads. These were the golden September days of plenty for the multitude of rodents, their harvest time of seeds.

A late-summer harvest of another kind was in full swing when we descended into the lower valleys of the wetter western side of the Great Divide. Colorado's famous peaches, the giant Hale and Elberta, were coming down from the orchard trees. We stopped at Paonia for half a bushel. So large they averaged only from twenty to sixty to a crate, they are covered with a somewhat denser fuzz than eastern peaches. Pickers frequently dust their necks and shoulders with starch or powder to avoid the discomforts of "peach fuzz itch." Almost two-thirds of the Colorado crop comes from an irrigated area about twenty miles wide and fifty miles long in the vicinity of Grand Junction. Here, where the Gunnison joins the Colorado, a "million-dollar wind" blows down the valley in the spring, reducing the danger of frost.

We were all enjoying the wonderful flavor of those giant peaches when we climbed onto the Black Mesa late that afternoon. Between Grizzly Ridge and Poison Springs Hill, we came to the north rim of the Gunnison's stupendous chasm. For millions of years, the mesa had risen slowly against the cutting flow of the river. Down through granite and gneiss and schist, through folds and veins and seams, down to some of the oldest rock on the face of the earth, the Gunnison had carved its way. In places the river narrows to a width of only forty feet and the sheer walls of the chasm rise to a height of more than 2,400 feet. To the Ute Indians the Black Canyon of the Gunnison was "The Place of High Rocks and Much Water." In 1933, land along ten of the canyon's fifty miles, including the spot where we stood, was set aside as a national monument.

It was about six o'clock when we reached the awesome edge of this abyss. Dusk was already gathering in its depths. The twisting gorge, with its broken side chasms, is gloomy, forbidding, a black canyon in reality. Peering gingerly over the brink, we caught, far, far below us, the faint glimmer of a white thread of foaming water. In the fading light we searched the canyon walls with our fieldglasses, hoping vainly to glimpse that sure-footed animal of the crags, the Rocky Mountain bighorn sheep. Along this canyon it has found one of its remaining strongholds in the West. By 6:30, the river had disappeared. Night was complete in the depths of the Black Canyon.

We stayed on in this remote and lonely place until the moon rose over the mesa. Its rays shone on the heavy masses of the shrub oak leaves, on the foliage of the mock orange and the mountain mahogany, on the twisted juniper and the ancient piñon pine—some seven centuries old—along the canyon rim. Somewhere, in these strange surroundings, we heard a familiar voice. A snowy tree cricket began its mellow song of the night.

With headlights switched on, we crossed the tableland and began the long, circuitous route back to the cabin on Spring Creek. Moths swirled into the beams of our lights. Half a dozen times, brilliant red eyes shone suddenly on the road before us and a resting Nuttall's poorwill shot up into flight. Once twin fawns bounded along the path of our illumination before they swerved away into the underbrush. And several times, along forty miles of dirt road that in wet weather becomes impassable, we caught sight of large porcupines waddling away into the surrounding darkness.

In the deep cleft of the Spring Creek valley, it was nearly eight o'clock each morning before the sun reached the log cabin. It had climbed well up in the sky, was almost overhead, when we started out with our friends on our last adventure in the region, a Jeep climb to a remote valley and Emerald Lake, that green gem among the little tarns of the Rockies.

Our road ascended toward Crested Butte through old mining country. It passed a gulch from which more than a quarter of a million dollars' worth of gold nuggets had been taken. Burros gone wild, relics of these early mining days, are still sometimes encountered in the remoter valleys. Once in our ascent we paused to examine with pleasure the richly red flower tubes of the fairy trumpets now in bloom. Ancient lichens sprawled across all the rocks around us. So slow is the development of some lichens in the high mountains of the West that half a century's growth can be measured in millimeters.

There was for years in Crested Butte an extraordinary fireplace decorated with living lichens of many colors. The local forest ranger selected each rock not only for its size and shape but for the form and color of the lichens it bore. After these stones were cemented together to form a large living room fireplace, the plants continued to thrive indoors. At regular intervals the rocks were moistened with a fine spray, the primitive plants absorbing as much as a quart of water at a time.

Our ascending way lifted us into a wide valley where the weathered buildings of an abandoned mining camp huddled together to form the Rocky Mountain Biological Laboratory. Here, each summer, scientists study problems relating to the natural history of the high Rockies. Now they were all gone, scattered among educational institutions across the United States, and the buildings stood silent and deserted. Beyond, climbing through virgin timber, we emerged suddenly with the vast panorama of the Elk Mountains spread out before us. Some of these jumbled peaks are still unnamed. Conditions there are comparable to those found by the mountain men when they first penetrated the Rockies.

Not long afterwards our dirt road dropped to its end. It dipped into a little valley set between steep slopes. Emerald Lake, its green-tinted water bordered by meadows strewn with mountain flowers, lay cupped at its bottom. That small, enchanted valley, as we looked down on it then, seemed to hold in concentrated form the near-at-hand beauty of the mountain heights.

Over us, as we wandered across the meadows later that afternoon, fine, shining threads of gossamer drifted. Even here, so high in the mountains, spiderlings were ballooning, riding the air currents, being dispersed widely on the winds of these late-summer days. In the Himalayas, Major R. W. G. Hingston found spiders on the naked rocks above vegetation, above all other forms of animal life. We remembered, as we watched this silken rain drifting by, that in older times gossamer was sometimes called "The Flying (or Departing) Summer."

All down the lower end of the valley, filling the deep gulch where the East River begins its flow toward a juncture with the Taylor and the formation of the Gunnison twenty-five miles away, a great bank of snow extended as far as we could see. It was no swiftly melting drift like the little snowfield below the rock slide. It was part of that phenomenon limited to high altitudes or high latitudes, the perennial or summer snow. In the dry northern Andes, not far from the equator, snowfields in summer begin at an elevation of about 20,000 feet. In Polar regions, they may descend almost to sea level. Many of the higher peaks of Colorado, like those of the Never Summer Mountains, have snow lines as well as timberlines. The former represents the lower limits of the perennial snow.

Oftentimes, depending on the dryness and the amount of sunshine received, one side of a peak will have a higher snow line than another side. In general, the climate of the region determines the lower limits of the snowfields. Over long stretches of time, this regional snow line may vary appreciably. Sometimes it slowly ascends the slopes; at other times it gradually drops lower. By recording such long-range shifts in snow line, scientists have found a yardstick for measuring changes in the local climate of the higher mountains.

That night, long after dark, as the Jeep was tilting and bouncing over the rocky road beside Spring Creek, Dwight shifted gears. There was a snap and a grind. We came to a helpless stop. Something in the internal anatomy of the clutch had sheared off. For sixty-five miles that day we had ridden through wild or sparsely settled mountain country. A breakdown in any one of a hundred places would have meant a long walk or a cold night spent among the peaks. But luck was with us. The accident happened almost within sight of the cabin door, hardly 200 yards away. Pushing the machine off the wheel tracks of the road, we soon had a fire going in the cabin stove and the smell of supper cooking filling the air.

When, at last, we went to bed for sound, refreshing mountain sleep, on this final night before bidding our friends goodbye, we could still see in our mind's eye that high valley with its emerald lake, its flower-filled meadows and its summer snow. In the morning we would leave for another world and for mountains of a different kind, mountains of sand, the highest and largest of America's inland dunes.

THIRTY-TWO

THE SAND BANNERS

WE WERE more than fifty miles away when we first caught sight of the low, shining, yellowish line drawn by the Great Sand Dunes along the base of the Sangre de Cristo mountains. The line grew wider, became more distinct, as we descended the broad San Luis Valley, the southernmost and the largest of the chain of five great "parks," or mountain-rimmed plateaus, that run from north to south down central Colorado. Gunnison lay seventy miles away over the crest of the Great Divide.

By now, almost all the weeks of the summer had, one by one, drifted away behind us—in the eastern mountains, along the Great Lakes, in the north woods, on the Great Plains. The autumnal equinox was drawing close. It was late in the afternoon of the summer season when we came to the shining dunes of this Colorado valley.

For ages, across the valley floor—flat, semi-arid, three times the size of the state of Delaware—the prevailing wind had swept out of the southwest. It had funneled away through Mosca, Medano and Music Passes, the lowest points in the immense wall of the Sangre de Cristos. Century after century, these winds, slowed suddenly on reaching the mountains, had deposited their burden of sand along a ten-mile stretch below the passes at the foot of the range. The ever-changing, wind-sculptured hills that thus came into being form the highest inland dunes of the United States. Here were no rolling, green Nebraska dunes. Here were naked peaks and hollows and curving ridges that lifted as high as 700 feet above the valley floor and swept in sinuous lines across miles of open sand. In 1932, eighty square miles of the area were set aside by presidential decree as the Great Sand Dunes National Monument.

Riding down the graded road that led us for twenty-three miles from the highway into the heart of the dunes, we saw Swainson's hawks perching on fencepost after fencepost. They were gathering together for the autumn migration. This hawk is one of the few raptors that move southward in flocks. Three other fenceposts, as we drew close to the dunes, were crowned with golden eagles. Each majestic bird, with deliberate, powerful down-strokes of wings that seemed enormous, rose heavily and sailed away. It headed directly for the dunes. Above this expanse of sun-heated sand, violent thermal currents rose far into the sky. One by one, the great birds curved into a corkscrew spiral and rose on soaring wings with the rising air. They climbed effortlessly up and up until they appeared no larger than sparrows under the drifting clouds. This portion of the sky directly above the dunes is the stairway, the escalator, of the golden eagles. On summer days they come from all the surrounding area to rise on the thermal updrafts. Lifted to great heights by this uprush of heated air, they head away, soaring mile after mile across the level valley.

On this day an east wind hurled itself through the passes and down the steep western slope of the Sangre de Cristos. It flung long, hard gusts across the sandhills and set all the dune tops smoking. Banners of sand streamed downwind from each upthrust peak. They extended—some for as much as 100 feet —as snow banners are drawn from the mountaintops in winter. They extended away like driven smoke, like a comet's tail, like sparks from an emery wheel. Each was darker, denser, narrower at the beginning; lighter, thinner, wider toward the end where the sand grains separated as they fell apart in a de-

THE SAND BANNERS

scending arc. So the wind for hours reshaped the dunes.

Most of all in a great wind do dunes come alive. Even in repose they rarely seem static. Even on the stillest days, cloud shadows riding up and down the slopes, parading across the flowing peaks and ridges, enhance the feeling of life and animation that the wind-shaped dunes impart. Watching the wind at work, watching it carrying away tongues and streamers of dune-top sand, watching it wearing down edges, sharpening points, watching the particles of rock drifting, piling up, flowing away, we were observing the creation of new dunes and the steady alteration of old ones. In these hours of wind and broken lighting and the slide of the cloud shadows, the whole area before us resembled moving, tumbling surf.

We walked out toward the edge of these dry breakers, pushed along, hurried by the gusts, the sand scudding all around us. The slopes, in the sunshine, had a toasted, gray-golden hue. When we scooped up a handful of the sand and examined it more closely through a pocket magnifying glass, we saw that the larger, tannish particles were roughly round while the smaller granules, more angular, appeared in a wide variety of colors—green, gray, white, pink, blue. In different illuminations, at dawn and sunset, at midday and in moonlight, even with shifts in the angle of the sun's rays, we found, delicate alterations in hue occur all across the flanks of the dunes. Among the sand grains, flour-fine flakes of gold have been discovered.

Moving about amid the gritty surf, we watched the sand sliding up the windward side of the hills and pluming or spilling over the crest. As it was built up on the lee side, the slope became steeper and steeper until it reached the "angle of repose," at which the pull of gravity and the inertia of the sand at rest were balanced. For dry material, this angle ranges up to thirty-four degrees. Additional sand slides down the steep slope. Eventually the "slip face," or profile of the lee side, becomes almost a straight line descending at the angle of repose. An immense sand dune, in the course of time, moves ahead, under the force of the prevailing wind, in a rolling or rotary manner, its substance turning like a loosely-held-together wheel. The windward sand is blown over the top of the dune and covered with more sand until, in the end, it once more becomes the windward sand.

At each gust, the grains about our feet began moving in unison, crawling like unnumbered minute insects around us. The tiny squeaking notes of their rubbing, infinitely multiplied, reached our ears in rising volume. This was the voice of the dunes, the shrill, high, thin song of the moving sand. Once a monarch butterfly scudded past on the wind, up and down, over the ridges and into the hollows beyond. For a long time, with our backs to the gusts, we stood close to the lip of a ridge where veils of sand lengthened and shortened with the varying strength of the wind. Each time we struggled along a ridge top we would stagger down into some protected hollow, pummeled and sandpapered by the flying grains. In a number of these hollows we found more than protection from the gusts. We found ourselves in the presence of a remarkable plant, one of the very few able to root themselves in this unstable sea of sand.

This was the lemonweed, a low-lying legume of the genus *Psoralea*. A member of the second largest group of flowering plants in the world, the pea family—its approximately 13,000 species exceeded only by the nearly 20,000 species of the sunflower family—the lemonweed produces its leaves along a running, perennial root. In descending into one small hollow, I trod upon one of the plants. The whole depression was filled with the rich scent of lemons. We examined a leaf, holding it up to the light. Translucent glandular dots were scattered all across it. In these glands the lemon-scented oils originate.

Where the wind had undercut them, the long roots lay exposed like lengths of yellow cord or binding twine. I paced off one portion of a root that lay at the surface. It measured

THE SAND BANNERS 329

eighteen feet. How far the rest extended underground I do not know. Successful dry-country plants are characterized often by abnormally long roots. One African species, the *Welwitschia*, a plant only twelve inches high, has roots that grow to a length of sixty feet during the short rainy season. Each of the roots we examined was dry and tough and hard. In the days before the white men, Indians of the region used such roots in making bridles for the half-wild horses they rode.

Our attention was caught by a blue-black ground beetle winding among the maze of stems of one clump of lemonweed. Then we forgot the beetle. We became engrossed in the sagacious search of a small dune-country caterpillar. Only about three-quarters of an inch long, it was green-hued and almost hairless. Its color closely approximated that of the lemonweed leaves on which it had been feeding. When first we caught sight of it, it was crawling rapidly away out across the loose sand. Its time for eating had ended; it was now in its period of "walking." Hurrying over the shifting, unsteady sand on this late-summer day, it was seeking a place to pupate. As we watched it, tongues of wind tumbled it over or swept the surface sand from under it. Each time it righted itself and pressed forward again, humping along in haste.

We followed its wandering course across a dozen feet or more of shifting sand. What hiding place was it looking for? There are, in these dunes, little areas where slightly larger and heavier grains of rock have collected together in what appear to be sandbars on the sand. The tiny stones are less easily moved by the wind. Such areas are weighted and relatively more stable. Not once did the caterpillar make any move to burrow into the loose sand. But as soon as it reached one of these patches, almost immediately it nosed down and began digging into the ground. In a surprisingly short time it was out of sight. Unseen, it would alter into its pupal form, spending the winter inert and helpless beneath the surface. Its act of instinctive wisdom had reduced the chances that it would be exposed by the wind and left for some bird or rodent to find. By choosing this slightly more stationary area in preference to the finer, more easily shifted sand, it tipped the scales toward the side of its own survival.

The following day when we returned again, the wind was gone and the unclouded sun was pouring its light down on the valley and the dunes were lifting in shining, fluid lines against the vast backdrop of the Sangre de Cristos. This range, representing some of the youngest mountains in the Rockies, contains eight peaks that rise above 14,000 feet. Along portions of this titanic eastern wall of the San Luis Valley, the forest runs in a wide band. It is bounded by a double tree line—above by the timberline of the heights, below by the dryness of the semi-arid valley floor that ascends part way up the slope.

The first American to cross the Sangre de Cristo range and visit the Great Sand Dunes was Lieut. Zebulon Montgomery Pike. On January 28, 1807, a little more than two months after he had discovered the peak that bears his name, Pike led his government exploring party down the western slope of the mountains and out among the dunes. Scanning the valley through a telescope from the highest peak of sand, he caught sight of the Rio Grande River where it curves to the south to flow past the present city of Alamosa, thirty-five miles away. Before Pike, no doubt, the roving Spaniards had reached the valley of the dunes. And before them—long, long before—the area had formed the home of ancient men whose Folsom-type artifacts are sometimes uncovered by the drifting of the sand.

As evening came on, that day, the wind freshened. Once more it poured down from the mountain heights. Clouds plunged over the saddle of Mosca Pass like foaming water over the lip of a dam. They rolled in a tumbling waterfall of vapor down the slope. But the white cascade descended only part way. It vanished, its moisture swiftly dissipating, when it reached the lower, warmer and drier layers of the air.

THE SAND BANNERS

Far below this cloud Niagara, we watched the many-skeined rivulets of Medano Creek catch and reflect the evening light. They appeared and disappeared, intertwining endlessly, to be lost at last beneath the sand. The gale, rushing down the mountainside, hurled itself through the cottonwoods beside the creek and plunged out across the dunes, adding to the silhouette of each sharply etched top the silhouette of a streaming banner of sand.

At 8,000 feet, a chill comes quickly with the falling of September nights, and we slipped on jackets before we started out with flashlights. It was now after the tourist season. We were alone, all alone, in the strangeness of the wind-torn night. Some years, kangaroo rats—those long-tailed, dry-country rodents whose powerful hind legs can hurl their five-inch bodies for as much as eight feet over the sand—are plentiful along the edge of the dunes. The fairy lace of their tracks runs everywhere. Over the sand, in and out among the shrubbery, we swung the beams of our torches in search of them.

Behind us the moon, slightly past full—the last full moon of summer—rose above the wall of the Sangre de Cristos. Out on the floor of the valley, the tips of the highest dunes were bathed in moonlight first. The pale illumination spread downward over the hills of sand while we still stood in the dark, untouched by the rays. Nothing is harsh or angular in moonlight. All is smoothed and softened, as moss softens the granite boulder. The dunes lay like faintly luminous clouds, their flowing lines imparting subtle variations to the lighting. We stood alone amid untamed and eerie beauty. Above there was the wildness of the sky, below the wildness of the sand. And all around us there was the wildness of the wind.

On a night like this it is easy to believe in an old legend of the region, the tale of the web-footed horses, animals with widespread hoofs that supported them on the soft sand. In spectral herds, their manes flying like the manes of the sand, they were thought to gallop across the dunes on nights when the moon was bright. How or when this legend started nobody knows. Perhaps it had its roots in moonlit glimpses of real animals, wild horses, the descendants of the steeds that escaped from the Spaniards. Once such creatures roamed, swift and free and unshod, all through the valley.

We walked for a time amid the nearer dunes, our moon-cast shadows stretching before us on the sand. Then we turned back again to the shelter of the cottonwoods, probing little glades, plumbing tangles with the moving beams of our flashlights. It apparently was a low point in the cycle of the kangaroo rats, for only once did we glimpse one of these leaping rodents. Only twice did we come upon their holes, each with a fan of delicate tracks extending away from the entrance. Small tan-colored moths gyrated in the beams. In one tangle our lights glinted on the shimmering wings of a sleeping dragonfly, and nearby we came upon a little pool, no more than three feet across, the path of the moon cutting its dark surface and little aquatic beetles turning and twisting in the water.

The roaring of the wind-surf was always around us. Violent gusts slashed at the cottonwoods. Under their tossing limbs, I could see the wandering fire of Nellie's torch threading among the grasses, running up and down the gray-green leaves of the yuccas, winding in and out among the dead branches of the fallen trees. For more than two hours we continued enjoying the fun of flashlight hunting. In this manner, amid the great gusts, beneath the last full moon of the season, beside the ever-changing hills crowned with their plumes of flying sand, we came to the end of this memorable latter day in our summer journey.

THE GREAT SAND DUNES with lemonweed growing in a hollow. Below, a stream disappears under the dunes.

SHADOW of Pike's Peak stretches east in the last sunset of summer until its tip reaches the horizon.

THIRTY-THREE

LAND OF THE FIVE SUMMERS

THE metal wings of a Cessna monoplane lifted us to 15,000 feet and carried us on a 150-mile swing over the plains and mountains. We took off from Peterson Field, east of Colorado Springs, with Henry Jend at the controls. At first the yellow plains spread beneath us, dotted with the tiny clearings of the harvester ants, the sand of every dry stream bed a lacework of cattle tracks. Then we were winging above the uptilted reddish rocks of the Garden of the Gods and past the convoluted road that climbs laboriously by switchbacks to the summit of Pike's Peak. Beyond we were among the high valleys and jagged mountains of the Rockies. Cripple Creek passed by below, all the surrounding slopes pockmarked, as by a prairie dog town, with the burrowings of prospectors. When we tilted into a wide turn toward home, the chasm of the Royal Gorge, with the greenish, silver-flecked thread of the Arkansas River running through it, lay directly below.

In this sweeping panorama, our sky view encompassed the infinite variety inhabited by the plant and animal life of this mountain state that is twice the size of England and six times as big as Switzerland. It is a land of five summers, of five life zones, where the season arrives and departs under different conditions.

Less than 200 miles from the southwestern corner of Colorado—where its boundary joins those of Utah, Arizona and New Mexico to form the only place in America where it is possible to stand in four states at the same time—the concept of life zones, each with its characteristic plants and animals, had its origin. It was on San Francisco Mountain, north of Flagstaff, Arizona, in the 1890's, that Dr. Clinton Hart Merriam made his basic studies.

In Colorado, five of Merriam's six life zones ascend from the plains to the mountaintops. The lowest, the Upper Sonoran or plains zone, ranges roughly from 4,000 to 6,000 feet in elevation. Above it the Transition or foothills zone extends from about 6,000 to 8,000 feet. Next comes the Canadian zone, running from about 8,000 to 10,000 feet. The Hudsonian zone reaches from 10,000 feet to timberline. Finally the Arctic or Alpine zone embraces the world of tundra and bare rocks above timberline on the higher peaks.

During the last days of our summer journey, swiftly passing days filled with endless change, we swung in another circle, a circle by land, a great circle of the mountain portion of this dramatic and wildly beautiful state. We saw close at hand the variety we had glimpsed in that eagle's-eye view from the air. Winding through valleys, climbing mountain highroads, we were always amid signs of the ebbing season. Each in its own way, each under the conditions of a different zone, the five summers of Colorado were drawing to a close.

Out on the high plains, in the Upper Sonoran, before we climbed into the mountains, we saw the end of summer coming with miles of dry sunflowers. Fleecy clouds, pretty clouds, clouds that never rained, drifted overhead. Killdeer called from the melon fields of Rocky Ford, in the south, and, in the north, "stubble ducks," the local name for ring-necked pheasants, stalked across the open land near Brush. The rattlesnake grass, *Bromus brizaeformis*, now with dry spikelets and inflated lemmas, stood ready to give off a startling buzz when brushed against by a passer-by. Beyond blooming time now were the rose-purple Rocky Mountain bee plant and most of the other nearly 500 wildflowers that grow on the

Colorado plains. But still shining out along the dry roadsides were the large white flowers of the prickly poppy.

The sweetness of summer, for me, becomes more concentrated toward the end. That wish of childhood, the longing to hold back the season, revives again. Somehow, as John Kieran observes in his *A Natural History of New York City*, it always seems that spring comes slowly but summer departs in haste. Wherever we went in our final wheeling, zigzag course around the state, that departure, day by day, seemed to accelerate its pace.

Climbing into the Transition zone and leaving the first of the five summers behind, we wound along a long, narrow valley west of the Royal Gorge. Extending almost due east and west, it was the site of a curious double summer, a precocious season and a retarded one. On the northern slope, first reached by the sun, plants had run their course and gone to seed by the time other plants of the same species, on the shaded opposite side of the valley, had just reached the peak of their blooming time. Here, only two or three hundred yards apart, we saw two versions of the same season drawing to a close.

When we mounted into the Transition zone we were entering the home of one of the most beautiful of all the world's beautiful trees—the Colorado blue spruce. It is found mainly along stream banks at elevations of between 6,500 and 8,500 feet. A waxy coating, in the form of a blue powder or bloom, covers the stiff, square and sharply pointed needles and gives the tree its color. Once, on a winding descent, we stopped entranced by the delicate hue of one of these conifers. From top to bottom it seemed bathed in pale-blue and silvery moonlight. This spruce is scattered. It is never found in dense stands or widespread forests. Its branches appear layered like those of a white pine. It has a many-storied beauty. New growth on the high Engelmann spruce of timberline often has a bluish cast. But the branches of these trees appear more haphazard in growth. They lack the layered effect of the blue spruce of lower elevations.

Another moment of beauty in the departing days of this second of the five summers came somewhere in the region of Pagosa Springs. We were not far from the New Mexican line, running west from Alamosa on the Navajo Trail. On our left the land dropped away in a long decline, from end to end a sheet of golden flowers. Hardly a foot high, hairy of stem and brilliant yellow of bloom, slender, branched annuals grew close together. Variously known as the goldenweed, the goldenbush, Jimmyweed and yellow daisy, this member of the sunflower family, with the scientific name of *Aplopappus*—sometimes spelled *Haplopappus*—ranges from Mexico northward across the dry country of the Southwest, ascending mountainsides as high as 9,000 feet. The blooming period of this plant is exceptionally long. At various stages of its range, it is in flower all the way from February to November, missing by only two months making a complete circle of the year.

But this slope was the exception. We had come to the time when, as Thomas Hardy expressed it in one of his Wessex poems, "summer's green wonderwork faltered." Brown leaves replaced the green. Where there had been flowers, now there were seeds. The wind, at every gust, whirled away the fluff of the fireweed. The world of plants was looking forward to another year. Everywhere the season grew visibly older. Change was in the crisper air. We, ourselves, were changing, too. Our systems behave differently at different seasons. We are more resistant to certain diseases at one time of year than at another. Our whole bodies respond to the seasonal shifts around us. In summer, for example, the iodine content of our bloodstream is greater than it is in winter.

About Labor Day, a human migration had begun in the mountains as vacationists returned to their winter habitats. Now, in the last stages of this movement, cars rushed past us, piled high with odds and ends from summer cabins. Trailers were on the move, heading south. Long red trucks pulled away from ranches, carrying cattle to market. The summer holidays were over; it was a workaday world again.

Once a whole circus went by, loaded on trucks and moving to winter quarters, its season on the road at an end. In the windows of stores along the way fall styles had appeared. Yellow school buses were now on the highway, and when we ran through small Colorado towns in the late afternoon we passed the crimson and gold and blue and maroon of high-school football teams scrimmaging in the first practices of the year. Many things, wild and human alike, pointed to the lateness of the season's hour.

When Hannibal marched his elephants over the Alps, the pass he is thought to have used was only 6,102 feet high. Many passes among the Rockies have an elevation of more than 10,000 feet. It was when we climbed toward such summits or ascended to the top of the higher mesas that we entered the third of Colorado's five summers, the realm of the Canadian zone. West of Walsenburg, at the top of the 9,382-foot La Veta Pass, we found ourselves surrounded by steep slopes clad in a coat of many colors, the autumn foliage of dense chaparral. On other mountainsides, we were amid lodgepole pine or among aspens, a few already clad in gold. In the northern part of this circular swing around the state, we saw, here and there, an aspen almost bare at the top. Leaf fall, that gradual and beautiful rite that brings the summer to a close and ushers in the autumn, had begun in the highlands.

The two most famous of Colorado's many mesas, or high tablelands, Mesa Verde, in the southwestern corner, and Grand Mesa, near the Utah border in the western portion of the state, represent the two extremes of the Canadian zone. Mesa Verde, with its ancient cliff dwellings, its skeleton weed and Apache plume—where we saw our first green-tailed towhee and dusky grouse and pinyon jay—is on the lower borderline, some in the Transition, some in the Canadian zone.

Grand Mesa, which we reached by a tortuous climb, is at the upper limits of the zone. When we attained the summit of this "world's largest flat-topped mountain," we found its high meadows, its coniferous forests, its more than 200 little lakes, some rimmed with waterlilies, all lying outspread in the brilliance of the late-summer sun.

The life zones, the varied summers of the mountainsides, shifted endlessly as we followed the up-and-down highways that led us west, then north, then east again. We rode across purple-sage highlands into Utah's uranium country where the sandblast of grit-laden winds had eroded rocks into eerie shapes. We skirted the Book Cliffs near Rifle, traversed the Axial Basin and followed the weathered gray-yellow rock of the Grand Hogback north to Meeker. Whenever we went north and whenever we went higher, we entered a later season. There the summer was running out more swiftly. At the foot of the Front Range, at an altitude of 6,336 feet, Manitou Springs has a growing season, between the earliest and latest killing frosts, of four and a half months. At the top of nearby Mt. Garfield, where the elevation is 12,467 feet, the growing season is only two months.

At times, in reaching the summit of the higher passes—Monarch, Wolf Creek, Red Mountain, Berthoud and Tennessee—we climbed above 10,000 feet and entered the Hudsonian zone. There, in the realm of the Alpine fir, the limber pine and the Engelmann spruce, we saw the fourth of these mountain summers nearing its end.

On our way to Leadville, the 10,200-foot "city in the clouds" with a climate that is described as "ten months of winter and two months mighty late in fall," we were close to the top of Tennessee Pass on the Continental Divide when we came upon a remarkable chain of beaver ponds. Forty or fifty of them descended like steps in a long water stairway down a steeply tilted little valley. Each pond was about seventy-five feet across and wider than it was long. Symbolic of the season's ending, on each little pond floated one or more pintail ducks, resting during their high-country migration. This chain of beaver dams not only provided miniature duck refuges, they also formed effective erosion and flood control. Another aspect of the work of those wild conservationists, the beavers, is evident throughout the Rocky Mountains. Their ponds silt up in time and thus produce the rich soil of many a mountain meadow. Ecologists have estimated that nine-tenths of the small meadows of the Rockies were produced by the filling in of beaver ponds.

Southward from Leadville, down the eastern slope of the Great Divide, we followed the earliest wanderings of the Arkansas River. A rushing mountain torrent here, it was far removed from the slow serpentine of the stream that I knew at Wichita, winding across the level land over its sandy bed between willows and cottonwoods. Foaming downward over rocks on its descent from Leadville to the Royal Gorge, it traces a twisting course through a legendary region, the land made famous by the gold rush of 1859.

Somewhere along the way, when we were well past the 18,000-mile mark for our trip and were nearing Florissant and Pike's Peak once more, we stopped for lunch at a small cafe. On the wall beside us as we ate, a gadget carried the sign:

"SWAMI KNOWS ALL. Drop one cent in the slot and ask any question. Swami will give you the answer."

Nellie, for fun, dropped in a penny and asked:

"Will we make a trip this summer?"

Out popped a printed card:

"Are you kidding?"

Only three of the great passes, Berthoud, Red Mountain and Monarch, each more than 11,000 feet high, carried us close to the topmost, the fifth summer of the Alpine or Arctic zone that we had entered earlier along the Trail Ridge Road. Above the wind-timber of tree line, out across the gale-swept tundra and the mats of cushion plants, at little cloud lakes and among the mountaintop rocks tinged green with lichens, summer, too, was coming to an end. On these lofty passes we saw this highest of the seasons close at hand; but we hoped to see it even closer at the summit of Pike's Peak, 14,110 feet above sea level, on the final day of the season.

Headlines now spoke of "premature autumn weather," of "early snows," of drifts that had already blocked the Trail Ridge Road. Would the Pike's Peak Highway be open to the summit on that last day of summer? That was the recurring question in our minds as we concluded our final swing around this mountain state in which there is so much to see. We came down from the Great Divide, from the ridgepole of the Rockies, for a last time. We descended onto the level floor of South Park. We crossed Muley Gulch and Wilkerson Pass and came to Florissant and the valley of the fossil insects. Beyond, at Woodland Park, we found a cabin made of logs and set among ponderosa pines on a ridge that looked up at the towering summit of the peak.

Although there are twenty-seven mountains in Colorado that are higher, Pike's Peak is the best known of all. Its name is alliterative. It is easily remembered. It became a household word through the slogan of the gold rush: "Pike's Peak or bust!" The mountain itself stands out boldly. It is visible for a hundred miles. It rises directly from the plains with no other challenging peaks of comparable stature close at hand. It is frequently called the most famous mountain in the world. Like Niagara Falls, it is an American synonym for the superlative.

From our cabin among the pines we gazed up at the summit. Our long and happy road through the second season of the year was almost ended. On that lofty eyrie overlooking the plains we hoped to spend the final day. When we went to bed that night, the season's end was hardly more than twenty-four hours away. In all our wanderings through Colorado, a state we remember with affection and awe, there had been a touch, a tinging of regret. For it was here, we knew, that our journey into summer would end.

THIRTY-FOUR

THE GREAT EYRIE—SHADOWS TO THE EAST

ONE o'clock. Snow is scudding past our cabin windows in the darkness. Five o'clock. The sky is clear and stars are shining. Seven o'clock. In the morning light, the great mountain towers above us, white-crowned with snow. Eight o'clock. The morning papers headline the weather news. All over the Rockies snow had fallen during the night. The Pike's Peak Highway is blocked by drifts at its upper end. Officials estimate it will take at least twenty-four hours to clear the road. Ten o'clock. We gaze at the mountaintop, glistening in the sun. We look at it through our fieldglasses. Our hopes of attaining the summit on this last day of our summer journey are gone. Eleven o'clock. We switch on the car radio. A news announcer reports that clearing operations have been speeded up. Officials have revised their estimates. The highway to the summit will be open by early afternoon.

It was two o'clock when we started up. Twenty miles of climbing loops and zigzags, switchbacks and 160-degree turns lay ahead of us. The spectacular road of graded dirt—built in 1915—carried us through four life zones. Part way up, it brought us out on top of the Front Range and we saw, 125 miles away to the west, the snow-mantled peaks of the Continental Divide. Just below timberline, seven miles from the summit, at an elevation of 11,525 feet, we stopped for a rest

THE GREAT EYRIE

at the Glen Cove parking field. Ten minutes later we were climbing again.

We left the last stunted Engelmann spruce behind and crept in zigzags up across the snow-covered boulder fields above tree line. Ravens were twisting and diving away in wild aerial games. There had been wind in the lower valleys, but here on the high slopes the air was almost still. A cloudless sky stretched above us and all the panorama that spread farther and farther away as we ascended was softened by the faint blue-tinted and shining haze. Our road extended before us, tilting upward, a plowed-out lane between recently installed snow poles that marked its wandering course among the huge pink-tinted boulders. This was one of the highway's last open days. In a week, or two weeks at most, it would be closed for the season. Snowdrifts twenty-five feet deep would cover it in winter.

We pulled up the last steep incline. Our wheels rolled on the level rock of the summit. We were 14,110 feet high, more than two and a half miles above the level of the sea, within fifty feet of being six times as high as the average altitude of the North American continent. At Tennessee Pass, we had climbed to 10,000 feet; at Monarch Pass to 11,000 feet; on the Trail Ridge Road to 12,000 feet. Here, at more than 14,000 feet, we were at the highest point of our summer travels. From this lofty eyrie, on this last day of the season, we could look away over the land through which we had wandered.

The last thing we expected to find on the naked shingles of the mountain roof was traffic congestion. Yet for a few minutes we seemed to be in Times Square. Thirty or forty other cars were parked or milling about—old Fords and new Fords, Cadillacs and Jaguars, Chevrolets and Volkswagens and Oldsmobiles, a Rolls-Royce and a Jeep. During the month of August alone, more than 20,000 motor cars had made the long ascent. They had transported in excess of 95,000 persons to the summit. The grand total for the summer season was well over a quarter of a million. Nobody knows how many human feet have walked about on the top of this mountain that Zebulon Pike believed no mortal man would ever climb.

More than half a century has passed since the first automobile reached the summit of Pike's Peak. On August 12, 1901, using the route of the cog railway, C. A. Yont and W. B. Felker drove a two-cylinder Locomobile steamer to the top. Now buses run up and down the mountain daily and a fleet of luxury limousines—air-conditioned, pressurized Cadillacs known locally as "Sherpa Caddys"—follow all the windings of the famed ascent. Each Labor Day, drivers from many parts of the world compete in a "Race for the Clouds," the winners maintaining for the long climb, with its switchbacks and hairpin turns, an average speed of more than a mile a minute.

Visitors to the mountaintop that day ranged from a baby carried about wrapped in a blanket to an elderly couple in their eighties. One woman in the low stone Summit House, which has weathered mountain gales since it was erected in 1882, was weighing herself to see if it was really true that she would weigh less at the top of a high mountain. Most of the cars stayed only a little while before they headed back down the road again.

As the afternoon advanced and the cars drifted away, we roamed over the boulder-strewn acres that form the flattened tip of the mountain. In all directions we looked away into the pastel shadings and smoky blues of the distance—toward the serrated line of the Great Divide, up the Front Range of the Rockies toward the north where the Trail Ridge Road now was blocked with snow, to the south and west where the Sangre de Cristos hid the Great Sand Dunes at their western base, to the east out over Colorado's level land, its tawny-yellow plains cut by the fine lines of branching roads and dotted with the far-spaced clusters of its towns and villages.

It was this eastward vista, viewed from this mountaintop at

sundown, that inspired Katharine Lee Bates' patriotic hymn, "America the Beautiful." Other nations have glorious mountains, breath-taking views. But none can approach the incomparable variety of wild beauty that is the heritage of this land —with its Yosemite, its Niagara, its Grand Canyon, its Old Faithful, its Carlsbad Cavern, its timberline gardens and deserts bursting into bloom, its Olympic forests and its ancient sequoias, its snow-clad peaks and all the teeming bird life of its coastal swamps. This gift of natural beauty is a rare and precious possession. The opportunity for citizens to enjoy it freely, now and in the future, is one of the inalienable rights of Americans.

As we stood discussing such things, two businessmen struck up an acquaintance nearby. They talked endlessly, loudly, always on the same subject: the clubs they had belonged to, how one had presided over a grand conclave, how the other had headed a committee that brought in twenty-two new members. All the while the great spiritual experience of the mountains was passing them by. Unseen, unfelt, unappreciated, the beauty of the land unfolded around them. The clubs of the world formed their world entire. It enclosed them like the home of a snail wherever they went. For them, the scene would have been just as moving if they had been hemmed in by billboards.

In time, the businessmen went down the mountain. One by one the cars rolled away. Nellie and I were almost alone in the stillness of the mountain heights. Far below we could see the tangled thread of the road we had climbed twisting across the boulder fields. Away to the north, through our glasses, we picked out the very ridge with its ponderosa pines where our log cabin waited with wood piled beside the fireplace. The rarefied air, in the late afternoon, grew swiftly cold. Even when, under the burning summer sun, the ground at the top of Pike's Peak reaches a temperature of 140 degrees F., the thin air five feet above it is only half as warm.

With the sinking of the sun, we stood by the tracks of the cog railway, east of the Summit House, and watched the shadow of the mountain creeping down the slope. Nineteen thousand miles away, as our course had carried us, we had seen the sunrise of the season's first day extend to the west the shadow of the New England mountains. Now the setting sun of the season's last day was casting to the east the shadow of this peak in the Rockies. Shadows to the west, shadows to the east, these enclosed our summer.

We watched the shade descend the slope, turning the snow blue among the pink rocks. We saw it grope its way among the foothills, darken the valley of Manitou Springs and the Garden of the Gods, extend out across the grid-work maze of the streets of Colorado Springs and then rush on, darker blue, over the blue plains beyond. At first its form was blunt and wide of base. Then it became more sharply triangular, ever elongating, its tip stretching out more keenly pointed. In a great spearhead, it plunged with accelerating speed across ranches and roads toward the far-off horizon. Pike's Peak rose like a titanic, 2½-mile-high sundial, its shadow, as we saw it, indicating the hour of the year, the lateness of the season.

A few minutes before six o'clock we watched the extended tip of this last of the summer shadows touch the horizon. For more than 100 miles it extended toward the east. Summer, the season of light, the season of life, was almost over. When the sun rose again it would stretch to the west the first shadows of autumn. We gazed in silence at the darkening land. Back there, in the direction the shadow finger pointed, lay the looping, twisting, back-tracking course of our glorious journey. There we had seen the painted trilliums and the mayfly storm, the falling stars and the dusty turtles, the ferns of Smuggler's Notch and the swallows of Niagara. As we watched, the shadow of the peak gradually lost its sharpness, faded slowly away, became swallowed up in the universal shadow of the mountain ranges. The sun had set on the sum-

THE GREAT EYRIE

mer season. All our wandering course now lay in the shadow of autumn's eve.

Still we lingered on. We were alone on the mountaintop, the last to leave. We remained while the dusk fell, while the wall of haze, touched with green, mounted along the eastern horizon, while all the higher peaks to the west were surrounded by a darkening saffron glow. Airway beacons twinkled along the lower mountaintops when finally we left the great eyrie and started down. We descended slowly, across the boulder fields, below the timberline with its ancient, contorted trees, so slow to grow, so slow to die, so slow to waste away. Winding downward, turning constantly, we descended into the blackness of the night.

Later, in the log cabin among the pines, we piled sticks of wood into the stone fireplace and sat in the warmth of the flames. Just so, ten years before, in that other cabin, now far away on the bank of the Pemigewasset in the White Mountains, we had sat by a fireplace during the last hours of the spring. There we had ended our first adventure with a season. Now from our windows we could see the peak of this western mountain lifting its gleaming upper slopes, white as alabaster, into the star-filled sky.

About ten o'clock, we stepped outside to listen to the sound of the wind in the long needles of the ponderosa pines. Smoke from our cabin chimney scented the air. To the northeast, a patch of glowing reddish light in the heavens caught our attention. It expanded as we watched, grew more brilliant. Then we noticed to the north another patch and still another. They swelled and merged and deepened their color. Behind them, pale bands of silver rose, probing upward toward the zenith. Unable to believe our good fortune, we stared at these resplendent northern lights, a red aurora.

For more than an hour this heavenly display continued with its spearing shafts of silver and its glowing pools of red. Their intensity waxed and waned, gradually fading away, then strengthening once more. At last the silvery streamers became too faint to see. Then the red flush paled and disappeared entirely. By eleven o'clock, the ethereal pageant was over. Entranced, we returned to the cabin, piled more wood on the coals and in the light of the crackling flames waited for the boundary minute of the season.

It came well after midnight, at 2:27 A.M. The fire had died down. We stood at the window looking up at the alabaster peak soaring aloft surrounded by stars. Each heavenly body, in the clear, cold mountain air, burned with a special brilliance, most with a blue intensity. Orion hung almost directly over the peak. The Milky Way flowed to its right. Here were the stars we had seen so often on our trip, the stars that bring, in their courses, all the seasons and that now were bringing the summer's end.

Half around the world, at a point above the Indian Ocean 500 miles west of the Maldive Islands, at that precise transition moment, the sun was shining directly down on the equator. Now the scales were tipped. As dawn and dusk are milestones in the roll of the earth, so the beginning of spring and autumn are milestones in its circle around the sun. Ever since the vernal equinox, the days had been longer than the nights. Now, at the end of summer, all through the fall and winter until March, nights would be longer than days. We turned away from the window, away from the glimmering of the mountain and the gleaming of the stars. The last day, the last hour, the last minute of the second season of the year were gone. In the silence of that star-filled night, the great summer of our lives had ended.

INDEX

Abandoned farms, 11
Abrasion by dust, 207
Academy of Science, French, 261
Adam's needle, 291
Adoption among birds, 89
"Adventitious bud" theory of bird's-eye maple, 135
Aeolian dust, 206
Agassiz, Alexander, 151
Agassiz, Lake, 161
Agassiz, Louis, 151, 286
Airplane ride over Colorado, 333
Alexander, Gordon, 310
Alien streets, 245
Allegan, Mich., 67
Alpena, Mich., 62
Alpine anemone, 309
Alpine fir, 338
Alpine flowers, 305-314
Alpine gold flower, 313
Alpine sandwort, 307
Alpine zone, 334
Alva, Okla., 269
"America the Beautiful," 345
American Eagle, The, 123
American Meteorite Museum, 259
American Ornithology, 29
American pipits, 312
America's many summers, 4
Amiel's *Journal*, 240
Andromeda nebula, 258
Androscoggin River, 7, 13
Anemone, Alpine, 309
Anemone, marsh, 99
Angle Inlet, Minn., 161, 164
"Angle of repose," 327

Animals
  handicapped, 101
  keep cool, 224
  suffer from flies, 165
Antelope, 227
Ants fight beetle, 245
Apache plume, 337
Apartment houses, bird, 284
Aplopappus, 336
Appalachian Trail, 7
"Apple tree" clew to bird's-eye maple, 134
Aquatic teeter-totter, 162
Arabian Nights, our own, 283
Arbutus, trailing, 99
Arctic plants, 305-314
Arctic three-toed woodpecker, 62
Arctic zone, 334
"Are you kidding?" 339
Aristotle, 53
Arkansas River, 240, 242, 281, 333, 339
Arrowheads, 166
Artemisia, 185
Artesian wells, 171
Aspen, 160
Aspirin, 244
Asplenium viride, 15
Auctions, 237
Auk, The, 123
Aurora borealis, 347
Au Sable River, 61
Automobile, first across America, 219
Automobiles as calendars, 238
Autumnal equinox, 348
Avens, 305

INDEX

Average elevation of North America, 220

Baby grebe, 177
Badger, 173, 192, 264
Bailey, Vernon, 55, 194
Bailey's Harbor, Wis., 97
Bait stands, 61
Bald eagle, 29, 122, 144
Balduf, W. V., 77
Bandit, encounter with, 246
Bank swallows, 160
"Barking squirrel," 188
Bartram, John, 30
Bat Cave, 233
Bates, Katharine Lee, 345
Battle in the dust, 245
Battleship Rock, 139
Bauer's hotel, 96
Bear Lodge Mountains, 186
Bear Mountain, 116
"Bear pits," 152
Bear River, 164
Bear with shattered jaw, 103
Bear watchers, 153
Bears, 154-156
  eat grasshoppers, 311
Beartooth Range, 113
"Beautiful Land," 232
Beauty, nature's use of, 76
Beaver ponds, 338
Beaverwood, 160
Beddoes, Thomas Lovell, 81
Bee-fertilized flowers, 309
Beetle vs. ants, 245
Beetle, ground, 329
Belt, Thomas, 148
Bent, Arthur Cleveland, 22, 89, 267
Berthoud Pass, 338
"Between hay and grass," 187
Bewick's wren, 265, 270
Bicknell's gray-cheeked thrush, 22
Big Dipper, 252
Big Horn Mountains, 171
Big Spring, the, 119
Big Traverse Bay, 162
Bighorn sheep, 321
Biology of Dragonflies, The, 227
Bionomics of Entomophagous Insects, 77
Bird apartment houses, 284

Birds
  adopt nestlings, 89
  danger tactics of, 218
  deaths of young, 92
  early migration of, 318
  first flight of, 167
  keep cool, 224
  killed by lightning, 123
  killed at Niagara Falls, 29
  one-track minds of, 86
  rivers of, 64
  skill in landing, 56, 267
  varied voices of, 175
Birds of the Lake Umbagog Region of Maine, 7
Bird's-eye maple, 130
Bird's Eye Veneer Co., 131
Bison, 71, 174, 182, 281, 298
Bistort, 305
Blackbird, handicapped, 102
Blackbird, red-winged, 319
Black Butte, 114
Black Canyon of the Gunnison, 319, 321
Black Canyon of the Gunnison National Monument, 321
Black Hills, 171, 186, 195, 202
"Black iron from heaven," 260
Black Mesa, 285
"Black stars," 260
Black tern, 177
Black walnuts, 66, 108
Black-backed woodpecker, 62
Blackbirds, red-winged, 319
Blackbirds, yellow-headed, 95
Black-tailed prairie dog, 193
Bladderworts, 166
Blaine Escarpment, 271
Blake, William, 7
Blue heron, 95
Blue jay, 8
  with injured wing, 104
Blue spruce, Colorado, 335
Bluebird, 8
Blueberries, 127
Bobolink, 14, 101, 176
Bogert, Charles M., 288
Bois d'arc, 275
Bonaparte, Napoleon, 247
Book Cliffs, 338
"Boomerang business," 142

INDEX

Box elders, 66
Box turtle, ornate, 285-289
Brandom, Jerry, 114
Breakdown, 324
Breakfast
  at midday, 180
  wayfarer's, 62
Brennan, L. A., 286
Brewster, William, 7
Bridger, Jim, 184
Britton Hill, 116
Broad-tailed hummingbirds, 318
"Broadway butterfly," 28
Brockway Mountain, 152
Broley, Charles L., 123
Bromus Brizaeformis, 334
Brown, Barnum, 182
Brown creepers, 318
Buck, John Bonner, 80
Buffalo, 71, 174, 182, 281, 298
Buffalo gourd, 285
Buffalo hunts, 182
Buffalo rocks, 174
Buffalo stampede, 182
Bunny-in-the-Grass, 122
Bunting, lark, 182
Bunyan, Paul, 119
Burnt Plains, 151
Bur reeds, 166
Burros gone wild, 322
Burrow of prairie dog, 193
Burrowing owl and prairie dog, 194
Businessmen on Pike's Peak, 345
Butterflies
  number of species in U.S., 9
  valley of, 265
Butterfly
  fossil, 301
  red admiral, 28, 45
  white admiral, 109

Cactus-juice milkshake, 108
Caddis flies, 12
Cahalane, Victor, 54
Call of prairie dog, 190, 191
Calosoma, 245
Calumet and Hecla Consolidated Copper Co., 151
Campbellsport, Wis., 96
Canada geese, 122
  voices of, 175

Canadian Field Naturalist, The, 90
"Canadian soldiers," 38
Canadian zone, 334
Canoe Voyage up the Minnay Sotor, 161
Canvasback, 176
Canyon wren, 202
Carbon Peak, 319
Carlyle, Thomas, 285
Carolina junco, 318
Carolina paroquet, 71
Cassiopeia, 252
Cathedral Mountain, 271
Cather, Willa, 234
Cattle in Nebraska, 211
Cave of the Winds, 31
Cecropia moth, 78
Cedar Point, 40
Cedar waxwing, 9
Census of trees, 274
Center of America, 221
Cessna monoplane, 333
Chain saw guides lost persons, 169
Changing seasons, 336
"Chant of the Whippoorwill," 129
Chaparral, 337
Charles Mound, 111
Cherokee, Okla., 268
Cherry orchards, 97
Cheshire cat, 270
Chestnut-collared longspurs, 174
Chestnut-sided warbler, 8, 109
Chickadee, 8
Chili coyote, 285
Chiming bells, 312
Chimney Rock, 218
Chionophila jamesi, 305
Chippewas, 139, 161
Chisholm Trail, 272
Chitin, 44
Chlorippe wilmattae, 301
Chuckwalla, 227
Cicadas, 181, 202, 265, 276
Cimarron River, 204, 270
Cinquefoil, rough, 120
Circle of the Seasons, 104
Circus, moving, 337
Cladophora glomerata, 30, 31
Cliff swallows, 319
Clinton, DeWitt, 29
Clintonia, 11, 17, 99

INDEX

Cloud shadow, 229
Cloudburst, 106, 249
Clover silk, 215
"Cock of the plains," 185
Cockerell, Theodore Dru Alison, 299
Cockerell, Wilmatte Porter, 300
Cold light, 77
Cold-blooded creatures, 222
Coleopterous fireworks, 125
Color in dreams, 186
Colorado blue spruce, 335
Colorado peaches, 320
Colorado Petrified Forest, 295
Colorado Springs, Colo., 294
Colorado, University of, 298, 300
Colson, Jake, 164
Columbine, 309
Comanche country, 290
Combines, harvester, 180
Comet hunter, 258
Comstock, John H., 33
Conch stew, 108
Concord River, 165
Condor, The, 104
Conestoga wagons, 218
Connecticut River, 4
Conrad, Joseph, 240
Continental Divide, 338, 342
Cooling off, 181
Copper Harbor, Mich., 151
Cormorants, double-crested, 123
Corn, 231-237
Cornish pasty, 121
Coronado, Francisco Vasquez de, 243, 292
Cotton grass, 19, 99
Coumarin, 51
Counting sheep, 210
Country fairs, 237
Courthouse Rock, 218
Courtship flight of scissor-tailed flycatcher, 265
Cowboy's delight, 285
Cow Pasture, the, 71
Cowper, William, 230
Cragen, Dr., 254
Cragin, F. W., 286
Crane fly, fossil, 302
Crane, sandhill, 127, 176
Crayfish bisque, 108
Creepers, brown, 318
Crepuscular creatures, 229
Crested Butte, 322
Cripple Creek, Colo., 294, 333
Crossbill, red, 299
Crossroads of life, 201
Crowfoot, 166
Crows, 319
Cruelty of horse-and-buggy age, 67
Cultivated fields, beauty of, 231
Curiosity, rareness of, 13
Cuscuta, 215
Custer, S. Dak., 207
Cutworms eaten by prairie dogs, 198

Daimonelix, 205
Dakota Artesian System, 171
Damage done by prairie dogs, 197
"Dark River," 142
Dartmoor, 6
Dating by automobiles, 238
Dawson Trail, 168
Days stolen from inevitable Time, 129
"Dazzle of the poplars," 160
Deadman's Gulch, 315
Deaths of young birds, 92
Deepest well, 264
Deer, 127
 feed on orchids, 100
 grace of, 12
 three-legged, 104
Deer, white-tailed, grow larger in the West, 174
Dentist-drill voice, woman with, 154
"Departing summer," 323
Destructive toe, man with, 136
Devil's corkscrew, 205
Devil's hair, 215
Devils Lake, 172
Devils Tower National Monument, 189
Dickcissels, 249
Dickens, Charles, 237
Dicoumarin, 51
Dinosaurs, 182, 205
Disintegration of mountains, 18
Distribution, mysteries of, 275
Dixon, James B., 104
Dodder, naked, 215, 284
Dog-mouse, 188
"Do he eat flowers?" 101

INDEX

Door Peninsula, 96
Double-crested cormorants, 123
Dove, rock, 31
Dragonflies, 172
Dreaming in color, 186
Drought, breaking of, 249
Dry-country plants, 329
Ducks, eclipse plumage of, 125
Dufresne, Frank, 103
Dunes, sand, 325-332
Durfee's Hill, 116
Dusky grouse, 337
Dust, abrasion by, 207
Dust devils, 204, 269, 279
Dust storm, 203
  one-field, 279
Dusty kitten, 262
Dusty sunset, 209
Dwarf lake iris, 99
Dwarf senicio, 309
Dwight, Timothy, 23

Eagle, bald, 29, 122, 144
Eagle, crippled, 104
Eagle, golden, 299, 326
Eagle Man, the, 123
Eared grebe, 173, 177
Earning a living, 94
Echo Lake, 45
Eclipse plumage of ducks, 125
"Eel flies," 38
Egret, 95
Elberta peaches, 320
Elderberry, 66
Elephant flower, 307
Elevation, average, of North America, 220
Elk Mountains, 322
Ellendale, N. Dak., 172
Emerald Lake, 322, 323
Enemies in nature, 213
Enemies of prairie dog, 191, 192
Engelmann spruce, 312, 335, 338
English sparrows, 276
Equinox, autumnal, 348
Erie, Lake, 35, 146
Eriogonum, 284
Eroded days, 241
Escanaba, Mich., 130, 131
Escanaba River, 119
Estes Park, 304

Estivation, 226
Eve's thread, 291
Exaggeration, humor of, 203
Experience, grains of, 201

Fabre, J. Henri, 166, 167
Face of the mountain, 18
Fairy dusters, 285
Fairy trumpets, 322
Falcon, prairie, 278
Fallon, Mont., 184
Farms, abandoned, 11
Fast growth of corn, 234
Fatigue, 269
Fawn, beauty of, 101
"Feast and fairy-dance of life," 47
Featherstonhaugh, George W., 161
Felker, W. B., 344
Fenceposts, stone, 273
Fenton, Mich., 59
Ferns of Smuggler's Notch, 15
Ferret, 192
Ferryboat navigates through mayfly storm, 48
Field Guide to the Butterflies, 9
Field mouse, 53-58
Fig newtons fed prairie dogs, 196
Finches, rosy, 311
Fir, Alpine, 338
Fireflies, 75-83
Fireplace, lichen, 322
First flight of birds, 166
First westward trip, 239
"Firsts," 63
"Fish flies," 38
Fish keep cool, 224
Fiske, John, 234
Flashlight hunting, 332
Flatiron Mesa, 318
Flaubert, Gustave, 157
Flesh flies, 202
Flicker adopts starlings, 89
Flicker nestlings adopted by Teales, 90
Flicker, red-shafted, 299
Flickers adopted by screech owl, 89
Flies
  annoy animals, 165
  on mountainside, 21
Floodlights at Niagara Falls, 27
Florissant, Colo., 294

INDEX

Flower flies in rough cinquefoil, 120
Flower Vase Rock, 139
Flowers, Alpine, 305-314
Flycatcher, scissor-tailed, 265, 275
Foam flower, 16
Following streams, 7
Folsom Man, 292
Food of sandhill cranes, 127
Foods, unusual, 108
Foothills zone, 334
Footpath in the Wilderness, 18
Ford, Model T, 238
Forest fire, 108
Fort Laramie, 184
Forty-fifth Parallel, 62
Fossil
    butterfly, 301
    crane fly, 302
    insects, 294-302
    largest, 296
    trees, 295
Fossils, superstitious dread of, 298
Four Corners, 333, 334
Franconia Notch, 1
Franklin's gull, 177
Freeman, Beryl, 288
French Academy of Science, 261
Friends University, 238, 240, 245
Frissell, Mount, 116
Front Range, 296, 338, 342
Frontiers of distribution, 275
Frye, John C., 206
Fuller, Albert M., 97
Furrows guide pioneers, 290

Gadwall, 176
Gallinules, 95
Garbage dumps as tourist attractions, 152
Garden of the Gods, 333
Garfield, Mount, 338
Geese, Canada, 122
    voices of, 175
Gentiana germanica, 308
Gentle hawks, 267
Geographical center of U.S., 221
Gerard, John, 98
Germfask, Mich., 121
Getting lost, 168
Getting the news from nature, 50
Giant silk moth, 28

Gila monster, 226
Ginger snaps fed prairie dogs, 196
Gitche Ganow, 137
Glaciers, grasshopper, 310
Glass Mountains, 271
Glendive, Mont., 180, 181
Goat Island, 34
Goats-milk fudge, 108
Goatweed butterflies, 265
Godwit, marbled, 176
Golden eagle, 299, 326
"Golden earth," 205
Golden plover, crippled, 102
"Golden River," 142
Golden Treasury, The, 240
Goldenbush, 336
Goldenrod, dwarfed, 306
Goldenweed, 336
Goldfinches, 8
Goldflower, Alpine, 313
Goldsmith, Oliver, 9, 10
Goldthread, 19
Goliath beetles, 44
Goose Eye Mountain, 7
Goose tansy, 120
"Goose-drowner," 106
Gophers, pocket, 314
Gore, Sir George, 184
Grain elevators, 278
Grains of experience, 201
Grand Hogback, 338
Grand Island, 132
Grand Junction, Colo., 320
Grand Marais, Mich., 137
Grand Marsh, the, 71
    draining of, 72
Grand Mesa, 337
Grand Portal, 139
Grand Tetons, 311
Granite Peak, 113
Grass fires, 211
Grass, how it survives cutting, 51
"Grass snipe," 278
Grasshopper glaciers, 310
Grasshopper Park, 314
Grasshopper winds, 310
Grasshoppers, 276
Gray, Donald V., 173
Gray squirrel, tailless, 103
Gray, Thomas, 130
Gray-headed juncos, 318

INDEX

Great Divide, 315
Great horned owl, 148
Great Lakes
  product of Ice Age, 60
  steppingstones, 137
Great Lakes region, rainfall in, 60
Great Salt Plains, 269
Great Sand Dunes National Monument, 325
Great Smoky Mountains, 202, 318
Grebe, baby, 177
Grebes, 173, 177
Green aphides, 166
Green Men, the, 50
Green Mountains, 14
Greenfork Top, 111
Greenhorn limestone, 273
Greensburg, Kans., 264
Green-tailed towhee, 337
Grinding stones, 10
Grizzly Ridge, 320
Ground beetle, 245, 329
Ground squirrels, 320
Grouse
  dusky, 337
  ruffed, 10
  sharp-tailed, 214
Guadalupe Peak, 114
Guide to Bird Finding West of the Mississippi, 266
Gulls of Hiawatha, 139
Gulls, waterlogged, 30
"Gully-washer," 106
Gunnison, Colo., 315
Gunnison, Black Canyon of the, 319, 321
Gunnison National Forest, 315
Gypsum, 271

Hailstorms, 279
Hairweed, 215
Hale peaches, 320
Halfaday Creek, 145
Hamilton, William J., 57
Handicapped creatures, 101
Hannibal, 337
Haplopappus, 336
Happy men, 150
Hardy, Thomas, 67, 152, 336
Harebells, 120
Harvest, wheat, 180

Harvester ants, 228
Harvester combines, 204
Haustoria, 215
Haviland, Kans., 254
Hawk, marsh, 311
Hawks, Swainson's, 326
"Hay meadows," 211
Hay used for paving and fuel, 211
Haying season, 67
"Hearing corn grow," 234
Heat, 268, 281, 291
Heat regulation, nature's, 222-229
Heat relief, 181
Heat wave, 180, 248
Hedge apple, 275
Heer, 296
Helium, 290
Henry, Alexander, 170
Henry, Cordia J., 121, 130, 142
Heraclitus, 51
Herball, Gerard's, 98
Hercules beetles, 44
Hermit thrush, song of, 23
Herons, 95
Herrick, Francis Hobart, 123
Hess, W. N., 80
Hexagenia affiliata, 39
Hexagenia limbata, 39
Hexagenia rigida, 39
Hiawatha country, 138
Hiawatha, gulls of, 139
Hiawatha National Forest, 138
Hide hunters, 183
High-bush cranberry, 10
Highest point of trip, 343
Highland, Western, 292
Highwayman, meeting with, 246
Hind in Richmond Park, A, 175
Hines, Bob, 102
Hingston, R. W. G., 323
History of Japan, 80
Hittites, 260
"Hobble skirt," clew to bird's-eye maple, 134
Hodges, Mrs. Hewlett, 259
Holboell's grebe, 173
Holdup man, 246
Holy Terror Reservoir, 318
Homeotherms, 222
Horicon marshes, 95
Horned larks, 278

356 INDEX

Horse-and-buggy age, cruelty of, 67
Horse with goggles, 293
Horse, passing of, 67
Horses, web-footed, 331
Hot day in Glendive, Mont., 181
"Howlers," 211
Hudson, W. H., 47, 175
Hudsonian godwit, crippled, 102
Hudsonian zone, 334
Hugo, Victor, 240
Human migration, 336
Hummingbirds, broad-tailed, 318
Humor of exaggeration, 203
Humphreys Peak, 114
Hunting by flashlight, 332
Hunting in Kankakee marsh, 72
Hunts, Red River, 183
Huron, Lake, 61
Hybridization of corn, 233

Ice-cream man, the, 214
Iliamna remota, 74
Indian Drum, 139
Indigo bunting, 109
Individuality, 210
Inevitable Time, 129
Inscription Rock, 40
Insect Pompeii, 295
Insects
    darker on mountains, 310
    "fall from sky," 310
    fossil, 294-302
    keep cool, 224
International Falls, 161
Inventing words, 63
Inverted Forest, 119
Iowa, the corn state, 232
Iris, dwarf lake, 99
Iris lacustris, 99
Iroquois "chant of the Whippoorwill," 129
Isle Royale, 151

Jack rabbit, 292
    and kitten, 262
Jack pine, 299
Jackson, H. Nelson, 219
Jacob's ladder, 309
Jail Rock, 218
Jay, Steller's, 299
Jaynes, Lockwood, 164

Jefferies, Richard, 129
Jend, Henry, 333
Jerimoth Hill, 116
Jimmyweed, 336
Joliet, Louis, 96
Josselyn, John, 10
Juncos, 318
"June flies," 38
Juneberry tree, 122
"Junebug fever," 37
Juniper, 321
    trailing or prostrate, 100
Juniperus horizontalis, 100

Kaempfer, Engelbert, 80
Kalm, Peter, 30
Kangaroo rat, 224, 225, 331, 332
Kangaroos, lassoing, 184
Kankakee River, 71
Kankakee State Park and Forest, 81
Kansas, windy, 279
Kansas City Star, 316
Katahdin, Mount, 110
Kebler Pass, 319
Keech, Clifford, 30
Keewaytin Ice Sheet, 96
Kelleys Island, 40
Kelso, Leon H., 197
Kempton, Berny, 184
Kerner, Anton von, 308, 313
Kettle moraine country, 96
Keweenaw Peninsula, 150
Killdeer, 83, 334
Kilvert, Francis, 160
Kimberly, Cora, 255
Kimberly, Eliza, 253
Kimberly, Frank, 253
Kimberly, Oren, 255
"King of the Currumpaw," 292
King, John A., 198
Kingbird, western, 179
Kingbirds, 166
King's crown, 306
Kiowa County, Kans., 253
Kirtland's warbler, 62
Kissimmee Prairie, 123
Kitchitikipi, the Big Spring, 119
Kites, Mississippi, 266-268
Kitten and jack rabbit, 262
Klots, Alexander B., 9
Knothole, world of a, 20

INDEX

357

La Creation, 41
Lady's-slippers, showy, 100
L'Aigle, France, 261
Lake of the Clouds, 157
"Lake Country"—Michigan, 60
"Lake flies," 38
Lake of the Woods, 161
Lakes of Minnesota, 159
Lamb, Wendell, 55
Lamprey, sea, 145
Land of Heart's Desire, The, 240
Land of Little Rain, The, 228
Land of lost lakes, 171
Land tortoise, 249
Langlois, Thomas Huxley, 39
L'Anse, Mich., 151
Largest fossil, 296
Largest well, 264
Lark bunting, 182
Larks, horned, 278
La Salle, de, Sieur Robert Cavelier, 70, 71
La Salle and the Discovery of the Great West, 70
Lassoing kangaroos, 184
"Last of the Beautiful," 237
Last day of summer, 348
Laughing Man, The, 240
La Veta Pass, 337
Leadville, Colo., 338
Leaping turtles, 288
Least flycatcher, 8
  call of, 109
Leatherleaf, 138
Leaves, number of, 50
Lee, Russell, 132
Leelanau Peninsula, 65
Lemonweed, 328
Leonid meteor shower, 261
Leopold, Aldo, 99
Letters on the Natural History and Internal Resources of the State of New York, 29
Level land, 277
Lewis and Clark, 170, 185, 188, 194, 292
Lewis, Meriwether, 292
Liberal, Kans., 283
Lichen fireplace, 322
Lichen, usnea, 318
Lichens, slow growth of, 322

Life
  abundance of, 47
  crossroads of, 201
  seeds of, 319
Life Histories of North American Birds of Prey, 267
Life Histories of North American Thrushes, Kinglets and Their Allies, 22
Life Histories of North American Woodpeckers, 89
Life zones, 333
Lightning, frequency of, 105
Limber pine, 338
Lime pie, 108
Limestone, Greenhorn, 273
Lincoln, Abraham, 68
Lindhuska, J. P., 57
Linnaea, 63
Linnaeus, Carolus, 9, 63, 78
List of Extreme and Mean Altitudes in the United States, 117
"Little dog," 188
Littlejohn, Flavius Josephus, 67
Lives of the Game Animals, 156
Living, earning a, 94
Living by sun time, 213
"Lizard's sun," 223
Lobo, the wolf, 292
Lobster rolls, 108
Loco weeds, eaten by prairie dogs, 198
Lodgepole pine, 299
Loess, 205
Long Trail, the, 18
Longfellow, Henry Wadsworth, 138
Longicorn beetle, 166
Longspurs, chestnut-collared, 174
Loon, 118
Loons, nuptial display of, 126
Loosestrife, 128
Lost lakes, 171
Lost in the woods, 168
Lott, William M., 89
"Louisiana marmot," 188
Love vine, 215
Lower Peninsula of Michigan, 60
Lower Souris National Wildlife Refuge, 171, 173
Lubbock, Tex., 195
Luciferase, 77

INDEX

Luciferin, 77
Luna moth, 78
Lunt, Myra, 246

Magazine Mountain, 114
Magnolia warbler, 9
Maid-of-the-Mist, 28, 30
Maize, history of, 233
Making a living, 94
Mallard, 176
    nonchalant, 35
Manganese mine, 318
Manistee River, 61
Manistique Swamp, 121
Manitou Springs, Colo., 338
Mansfield, Mount, 14, 17-23, 99
Mantzelia nuda, 213
Maple, bird's-eye, 130
Marbled godwit, 176
Marcellina, Mount, 320
Marcy, Mount, 110
Marmots, 309, 314
Marquette, Pere, 96
Marriage-year trip, 238
Marsh anemone, 99
Marsh hawk, 311
Martha, last passenger pigeon, 65
Martin, crippled, 101
Martins, 27
Mather, Cotton, 64
Mating dance of mayflies, 45
Mayfield, Harold, 62
Mayflies, 36-49, 119
Mayfly storm, 36
Meade, Kans., 206
Meadow lark, 181
    without a tail, 56
Meadow mouse, 53
"Meal you can carry in your pocket," 121
Meat-hunters, 72
Medano Creek, 331
Medano Pass, 325
Medicine Lodge River, 268
Meloidae, 226
Melons, Rocky Ford, 334
Memoranda, 130
Memories, 10
    of Wichita, 240-248
Merriam, Clinton Hart, 194, 334
Mesa Verde, 337

Messier, 258
Meteor Crater, 257
Meteor shower, 251
Meteorite Farm, 253
Meteorites, 252-261
Miamis, 71
Miceless House, the, 59
Michigan, Lake, 65, 119
Michigan, "Lake Country," 60
Michigan, Lower Peninsula of, 60
Michigan, Upper Peninsula of, 119, 137
Michigan vineyards, 66
Microclimates, 307
Microtus pennsylvanicus, 53
Midges, mating dance of, 178
Midnight Bandit, the, 246
Migrating birds, 318
Migration
    human, 336
    of sheep, 320
    vertical, 318
    of waterfowl, 176
Migratory flocks, 176
Mile of grasshoppers, 221
Miller, Gerrit S., Jr., 80
"Million-dollar wind," 320
Miner's candles, 285
Miner's Castle, 139
Miniature rose, 306
Minnesota, lakes of, 159
Mint cultivation, 73
Minted peas, 17
Mio, Mich., 62
Mirages, 186, 278
Misquah Hills, 112
Mississippi kites, 266-268
Mock orange, 321
Model T, 238
Modern life, complications of, 84
Moisture output of corn, 236
Moldenke, Harold, 51
Mole, star-nosed, 66
Momence Ledge, 74
Monarch Pass, 315, 338, 343
Montmartre, 271
Moonlight smooths all, 331
Moosewood, 9
Morgan, Lewis H., 151
"Mormon signposts," 183
Mosca Pass, 325, 330

INDEX

Mosquitoes, 148
Moss campion, 306-307
Moth, giant silk, 28
Mountain, face of, 18
Mountain mahogany, 321
Mountain plants have woolly stems, 313
Mountain ranches, 319
Mountain, walking, 319
Mountain weathers, 312
Mountain-top soil, 307
Mountains
   disintegration of, 18
   plants bloom earlier on, 308
Mourning dove, voice of, 175
Mourning warbler, 157
Mouse, field or meadow, 53-58
Mouse River, 170
"Mouseometer," 55
Movement of sand dunes, 328
Mud Lake, 151
Muma, A. R., 30
Munising, Mich., 139
Music Pass, 325
Muskeg Bay, 161
Muskmelons, Rocky Ford, 316
Muskrat, 95
My Antonia, 234
Mysterious Lake, 318

Naked dodder, 215
Names of places, picturesque, 317
Names of western plants, 285
Naming lakes, 160
Napoleon Bonaparte, 247
Nashville warbler, 8
Natural beauty of America, 345
Naturalist in Nicaragua, The, 148
Naturalist's pace, 6
Nature
   beauty and wonder in, 33
   enemies in, 213
   getting the news from, 50
   rhythms in, 78
Nature's temperature controls, 222-229
Nature's use of beauty, 76
Navajo Trail, 336
Needle Rock, 318
Negaunee, Mich., 151
Nellie, 2, 7, 16, 17, 19, 20, 59, 63, 67, 110, 125, 126, 157, 163, 178, 186, 209, 214, 238, 256, 309, 332
Nervous man, the, 140
Nesting sites, scarce, 284
Nestlings adopted, 89
Never Summer Mountains, 323
New England a guide to botany, 15
Niagara Falls, 25-35
Nicolet, Jean, 96, 107
Night heron, 95
Night-blooming campion, 120
Nighthawks, 237, 318
Nininger, Harvey H., 259
Niobrara River, 205
Noise Needers, the, 21
Nonchalant mallard, 35
North America, average elevation of, 220
North Truchas Peak, 115
Northern Highland Province, 117
Northern lights, 347
Northwest Angle, 161
Northwest Passage, 8
Notes, making, 130
"Nuisance spots," 258
Nuptial display of loons, 126
Nuthatch, pigmy, 299
Nuttall's poorwill, 321

Occupations, 94
Oeningen, Bavaria, 296
Oh-Be-Joyful Creek, 318
Ohio Pass, 319
Oil beetles, 226
Ojibway, 83, 139
Ojibway chant, 83
Oklahoma Panhandle, 284
Old Blue Mountain, 114
Old man of the mountain, 313
Old Speck Mountain, 7
Oliver Twist, 237
One-cloud rains, 217
One-field dust storm, 279
One-track minds of birds, 86
Orchids, 97-107
Oregon Trail, 217
Orion, 260
Ornate box turtle, 285-289
Ortenburger, A. I., 288
Osage orange, 274

INDEX

Osceola, 237
Oskaloosa, Iowa, 237
Otter, 127, 143
Otter slides, 143
Our Lord's candles, 291
Oven bird, 8
Owl, great horned, 148
Oxalis, 15

Pack rats, 316
Pageant of Summer, 129
Pagosa Springs, Colo., 336
Paintbrush, 306
Painted trillium, 19
"Pancake Hub of the Universe," 283
Panhandle, Oklahoma, 284
Panhandle, Texas, 290
Parkman, Francis, 70
"Parks" of Colorado, 325
Parnassia, 308
Parry's clover, 306
Parula warbler, 8
Passenger pigeon, 64-65, 71, 182
Paulson, Clint, 132
Pawpaws, 108
Peach-fuzz itch, 320
Peaches, Colorado, 320
Peas, minted, 17
Pecking order, ursine, 154
Pectoral sandpiper, 278
Pedicularis groenlandica, 307
Pelican, white, 176, 177
Pemigewasset River, 4, 347
Pennington, Dwight, 315
Pennington, Esther, 315
Penstemon, 314
Pepys, Samuel, 310
Perry, Oliver Hazard, 40
Perseids, 251
Persimmons, 108
Peshtigo River, 108
Peters, George H., 34, 109-117
Petrified trees, 295
Petroglyphs, 40
Pettingill, Olin Sewall, 266
Phalarope, Wilson's, 176, 212
Pheasants, ring-necked, 334
Phillips, Andrew J., 59
Phillips, Clifford J., 59
Phillpotts, Eden, 6
Phoebe, 8

Photinus pyralis, 77
Pictured Rocks, 139
Picturesque place-names, 317
Pied-billed grebe, 173
Pigeons, individuality of, 31
Pigeons at waterfalls, 31
Pigmy nuthatch, 299
Pike Petrified Forest, 295
Pike, Zebulon M., 281, 330, 344
Pike's Peak, 114, 294, 333, 339, 340, 342-347
Pike's Peak Highway, 342
Pincushion tree, 10
Pine, lodgepole, 299
Piñon pine, 321
Pintail, 176, 339
Piolated warbler, 312
Pioneer surveyors, 143
Pioneers guided by furrows, 290
Pipits, American, 312
Pipsissewa, 99
"Place of High Rocks," 321
Place-names, picturesque, 317
Plains zone, 334
Plants, Arctic, 305-314
Plants bloom earlier on mountains, 308
Plants, dry-country, 329
Plaster of Paris, 271
Platte River, 217
Plentywood, Mont., 179
Plover, snowy, 270
Plover, upland, 175
Plumed grass, 266
Pocket gophers, 314
"Pocket Pups," 276
Poetic names of western plants, 285
Poikilotherms, 222
Poison Springs Hill, 320
Poisoning of prairie dogs, 196
Pokagon, Simon, 68
Pokagon State Park, 69
Polemonium, 115
Polygala paucifolia, 99
Polyphemus moth, 28, 78
Pony Express, 218
Pope, Clifford H., 286
Poplars, "dazzle" of, 160
Popple, 160
Porcupine Cone, 318
Porcupine Mountains, 111, 157

INDEX

Porcupine on Mt. Mansfield, 20
"Porcupine nests," 100
Porcupines, 321
Portage Prairie, 71
Porte des Morts, 96
Potash towns, 214
Potawatami, 68, 71
Potholes, 10
Powder River, 184
Prairie Dog Town Fork, 188
Prairie dogs, 188-200
Prairie falcon, 278, 311
"Prairies of Louisiana," 200
Praying mantis kills field mouse, 54
Preparations for travel, 2
"Presto Pups," 40
Prevailing winds, 204
Pringle, Cyrus G., 15
Prostrate juniper, 100
Protection of prairie dog, 189
Psoralea, 328
Pull-down, 215
Pulpit Rock, 139

Quarterly Review of Biology, 80
Queen Mary, 131
Queen of the Woods, 68
Quinet, E., 41

Rabbit, snowshoe, 11, 66
Rabbits, influence on orchids, 99
Race for the Clouds, 344
Racers, 228
Radisson, Pierre Esprit, 139, 148
Rail, Virginia, 95
Rail, yellow, 124
"Rain follows the plow," 281
Rain, heaviest recorded, 106
Rainbows, 33
Raindrops, 106
Rainfall in Great Lakes region, 60
Rain-making, 281-283
Rains, one-cloud, 217
Rat Lake Fire Tower, 117
Raton Pass, 292, 293
Rats, kangaroo, 331, 332
Rattlesnake grass, 334
Rattlesnake and prairie dog, 194
Rattlesnake, sandhill, 213
Rau, Phil, 78
Raven roads, 159

Ravens, 152, 159, 343
Record, Samuel J., 135
Red admiral butterfly, 28, 45
Red Bank, Wis., 107
Red crossbill, 299
Red Mountain Pass, 338
Red paintbrush, 306
Red River hunts, 183
Red squirrel, 3
Red thorn apples, 108
Red-eyed vireo, 109
Redhead, 176
Red-necked grebe, 173
Red-shafted flicker, 299
Redstart, 8
Red-winged blackbird, 319
   one-legged, 102
Rehabilitating a river, 61
Relief from heat, 181
Remembered river, 241
Resolute, 161
Rhubarb-and-strawberry pie, 61
Rhythms in nature, 78
Rib Mountain, 117
Rifle River, 61
Ring around a lake, 65, 119
Ring-necked pheasants, 334
Rio Grande River, 330
River of two thousand bends, 71
River, upside-down, 270
River within a river, 164
Rivers of birds, 64
Rivers, nature's wilderness roads, 7
Rivers, rehabilitating, 61
Road construction, 205
Robbins Ditch, 74
Roberts, Kenneth, 8
Robin, 8, 179
Rock dove, 31
Rock glacier, 319
Rocky Ford muskmelons, 316, 334
Rocky Mountain bighorn sheep, 321
Rocky Mountain Biological Laboratory, 322
Rocky Mountain National Park, 304
Rocky Mountain snow willow, 306
Roosevelt, Theodore, 189
Rose crown, 306
Rose Lake Wildlife Experiment Station, 57
Ross Allen's Reptile Institute, 247

INDEX

Rosy finches, 311
Rough cinquefoil, flower flies in, 120
Rough-winged swallows, 27
Royal Gorge, 242, 333
Ruddy duck, 176
Ruffed grouse, 10
Rydberg, P. A., 306
Rydbergia grandiflora, 313

Sage hen, 184
Sagebrush, 185
St. Exupery, Antoine de, 32
St. Joseph River, 70
St. Lawrence River, 24, 137
Saline River, 277
Salmon chowder, 108
San Francisco Mountain, 334
San Juan Mountains, 319
San Luis Valley, 325
Sand dunes, 325-332
Sanders, Jack, 255
Sandhill crane, 127, 176
Sandhill rattlesnake, 213
Sandpiper, pectoral, 278
Sandusky, Ohio, 36
Sandwort, Alpine, 307
Sangre de Cristo Mountains, 325, 330
Santa Fe Trail, 285, 293
Sap of early summer, 14
Sassafras, 66, 108
Sault Ste. Marie, 138
Scarlet tanager, 109
Scaup, 176
Schlief, Emil, 123
Schmidt, Karl P., 289
Schoolcraft, Henry Rowe, 138
Scientific American, 198
Scissor-tailed flycatcher, 265, 275
Scotts Bluff, 217
Screech owl adopts flickers, 89
Scrophularia leporella, 122
Scudder, Samuel Hubbard, 295
Sea lamprey, 145
Seasons, The, 240, 248
Seasons, the, 2
 change of, 336
Second molt of mayflies, 43
Sedum integrifolium, 306
Seeds of life, 319
Selby, Gertrude, 26

Selby, James A., 26
Selene, 271
Selenite, 271
Seney National Wildlife Refuge, 121
Senicio, dwarf, 309
Seton, Ernest Thompson, 156, 292
Shadbush, 122
Shade, a lifesaver for animals, 227
Shadow Line, The, 240
Sharp-tailed grouse, 214
Sheep
 counting, 210
 on a hot day, 181
 migration of, 320
Sheep, bighorn, 321
Sheppard, Roy W., 27
"Sherpa Caddys," 344
Shinleaf, 99
Shore birds, 176
Short-tailed weasel, 66
Shoveler, 176
Showy lady's-slippers, 100
Shrikes, 278
"Shrubaceous," 63
Singleton Ditch, 74
Sioux, 161, 261
"Sip," a small swallow, 63
Skeleton weed, 337
Skipping stones, 157
"Sky pilot," 115
Slater, Ada, 145, 149
Slater, Ken, 142, 145, 149, 169
Slaughter of game, 72
"Sleepers and Weekers," 119
Sleeping Bear Dune, 65
"Slip face," 327
Smithsonian Institution, 219
Smuggler's Notch, 14-17
 ferns of, 15
Snow lines, 323, 324
Snow willow, 306
Snow-lover, 305
Snow-on-the-mountain, 265
Snowshoe rabbit, 11, 66
Snowy plover, 270
Snowy tree cricket, 321
Soap Basin, 318
Soap, yucca, 291
Soaproot, 291
Soapweed, 291
Social nature of prairie dog, 190

INDEX 363

Soil, mountain-top, 307
Soo Locks, 62
Sora, 95
Souris midges, 177
Souris River, 170
South Bend, Ind., 70
South Park, 315
South Truchas Peak, 115
Southwest as geology textbook, 15
Spanish bayonet, 291
Spanish dagger, 291
"Sparrow starvers," 276
Sparrow, white-crowned, 318
Sparrow, white-throated, 3, 12, 13, 23
Sparrows now follow cars, 276
Spear-by-moonlight, 159
Spearfish Canyon, 202
Sphinx moth, 309
Spiderlings, mountain, 323
Spiral trees, 316
Spleenwort, *asplenium viride*, 15
Spongy soil, 73
"Sportsman's paradise," 142
Spring Creek, 315
Spruce, Colorado blue, 335
Spruce, Engelmann, 312, 335, 338
Spruce Knob, 115
Sprunt, Alexander, Jr., 123
Squaw bush, 10
Squirrel, red, 3
Stairway of the golden eagles, 326
Staked Plains, 188
Stampede of buffalo, 182
Starlings
  adopted by flicker, 90
  blockaded by woodpecker, 84
  handicapped, 102, 103
Star-nosed mole, 66
Stars, 348
State birds, 238
Static electricity, 208
Steller's jay, 299
Steppingstones of Great Lakes, 137
Sterling Mountains, 14
Stone fenceposts, 273
Stonecrop, 306
Storer, John H., 53
Storm on Lake of the Woods, 163
Storm on the mountains, 312
Stormy Lake Superior, 140

Strangleweed, 215
Strawberry pickers, 66
Streams, following, 7
Strecker, John K., 286, 287
Structure of corn, 235
"Stubble ducks," 334
Sturgeon, 163
Sugarbush Hill, 109
Summer
  "icumen in," 24
  last day of, 348
  sap of early, 14
  season of life, 346
  a stable season, 2
  a static season, 230
  vacation time, 3
Summer solstice, 3
Summer suits of birds, 224
Summers, America's many, 4
"Summer's green wonderwork," 336
Sun god, 313
Sun time, 213
Sunday River, 7
Sunflowers, 277, 334
Sunset, dusty, 209
Superior, Lake, 137, 140
Superior National Forest, 112
Super-sundae, 140
Surveyors, early, 143
Swainson's hawks, 326
Swallows, bank, 160
Swallows, cliff, 319
"Swami knows all," 339
Swammerdam, Jan, 39
Swamp-candles, 128
Swampy Pass, 319
Swifts, 237
Swifts, white-throated, 202
"Swog," 63
Synchronized flashing of fireflies, 79

Tahquamenon River, 142
Tahquamenon Swamp, 141
Tailwork of Mississippi kites, 268
Tall tales, 203
Teal, 176, 212
Teales adopt flicker nestlings, 90
Teeter-totter, aquatic, 162
Temperature controls, nature's, 222-229
"Ten months of winter," 338

INDEX

Tennessee Pass, 338
Tepee Gulch, 318
Tern, 95
Tern, black, 177
Terns at Niagara Falls, 26
Terrapene ornata, 286
Terror Creek, 318
Tertiary Insects of North America, The, 295
"Texas birds of paradise," 265
Texas Panhandle, 290
The Ridges Sanctuary, 97
Theodorsen, Theodore, 280
Thimbleberries, 108
Thomas-Elizabeth, 246
Thomson, James, 240, 248
Thoreau, Henry D., 13, 34, 35, 101, 244, 305
Thorp, James, 194, 204
Thrush, Bicknell's gray-cheeked, 22
Thrush, hermit, 23
Thrush, wood, 3
Thrushes, voices of, 109
Thunderbird, 299
"Thunderstones," 260
Thunderstorms, 105
Tiarella cordifolia, 16
Tiger swallowtail, 9
Tillyard, R. J., 226
Timber wolves, 165
Timberline, level of, 313
Timbers of the New World, 135
Time, inevitable, 129
Time is the river, 241
Tincup Pass, 318
Toads, 248-250
Toadstool rocks, 207
Toft, Emma, 100
Toft's Point, 100
Tongue hunters, 183
Tornadoes, 279-281
Tortoise, land, 249
Tourists feed prairie dogs, 196
Towhee, green-tailed, 337
Tractors, 67
Traffic congestion on Pikes Peak, 343
Trail Ridge Road, 304, 339, 343
Trailing arbutus, 99
Trailing juniper, 100
Transactions of the Kansas Academy of Science, 286

Transcontinental automobile trip, first, 219
Transition zone, 334
Travels in New England and New York, 23
Traven, Olivia, 97
Treasure Chest of Michigan, 151
Tree census, 274
Treeless land, 274
Treeless plains, 289
Trees, spiral, 316
Triceratops, 182
Trillium, painted, 19
Turtle, ornate box, 285-289
Turtle steak, 108
Turtles of the United States and Canada, 286
Twilight of the Gods, 3
Twinflowers, 63
Tyrannosaurus rex, 182

Umbagog, Lake, 7
University of Colorado, 298, 300
Unusual foods, 108
"Unwearied wheel" of nature, 230
Upland plover, 175
Upper Peninsula of Michigan, 119, 137
Upper Sonoran Zone, 334
Upside-down river, 270
Ursine pecking order, 154
U.S. Fish and Wildlife Service, 145, 170
Usnea lichen, 318
Utes, 295, 304, 321

Vacationists, 336
Valentine National Wildlife Refuge, 212
Valley of the butterflies, 265
Valley of the fossil insects, 294-302
Valparaiso moraine, 72
Vanessa atalanta, 28
Veery, song of, 13
Vermont as fernland, 15
Vertical migration, 318
Viburnum opulus, 10
Victoria, Lake, 138
Vineyards in Michigan, 66

INDEX

Vireo, 8, 109
Virginia rail, 95
Visitors to Pike's Peak, 343, 344
Voice of sand dunes, 328
Voices of birds, 175
Vole, 53
Voyages of Pierre Esprit Radisson, 148

Wah-Wah-Taysee, 76
Walking Mountain, 319
Walk-in-the-Water, 40
Wall of round stones, 276
Wall Street Journal, 244
Wallace, Alfred Russel, 300
Wallace, George John, 22
Wand lily, 309
Warbler, chestnut-sided, 8, 109
Warbler, Kirtland's, 62
Warbler, Magnolia, 9
Warbler, Mourning, 157
Warbler, Nashville, 8
Warbler, parula, 8
Warbler, piolated, 312
Warm-blooded creatures, 222
Warroad, Minn., 161
Washington, Mount, 110
Water beech, 15
Water elder, 10
Water lilies, 166
Water-cooling system of kangaroo rat, 225
Waterfalls, pigeons at, 31
Waterfowl, migration of, 176
Waterlogged gulls, 30
Watermelon sherbet, 237
Watersmeet, Mich., 119
Wayfarer's breakfast, 62
Wealth from outer space, 253
Weasel, short-tailed, 66
Weather news, 342
Weather wizards, 282
Web-footed horses, 331
Webster, Daniel, 290
Well, deepest and largest, 264
Welland Canal, 145
Welwitschia, 329
Western End of Lake Erie and its Ecology, The, 39
Western grebe, 173
Western highland, 292

Western kingbird, 179
"What does *he* do?" 94
Wheat harvest, 180
Wheel of the year, 230
Wheeler Peak, 116
Whetstone Mountain, 318
Whippoorwill, 129
"Whip-poor-will's shoes," 100
Whipsnakes, 228
Whirlwinds, 279-281
Whistling swans, 29
White evening star, 213
White Mountains, 1
White pelican, 176, 177
"Whitecaps," 211
White-crowned sparrows, 318
White-tailed deer, grow larger in the West, 174
White-tailed prairie dog, 193
White-throated sparrow, 3, 12, 13, 23
Whiting, Adrian P., 123
Whitney, Mount, 110, 115
Wichita, Kans., memories of, 240-248
Wichita Mountains, 271
Wickham, H. F., 296
Widgeon, 176
Wild burros, 322
Wild flavors, 108
Wild rice pancakes, 108
Wild strawberries, 108
Wildcat's track, 10
Wilderness State Park, 63
"Willow bugs," 38
Willows, 244
Wilson, Alexander, 29, 64
Wilson's phalarope, 176, 212
Wind Cave National Park, 195, 198, 202
Wind, Sand and Stars, 32
Windbreaks, 204
Windmills, 212
Winds, prevailing, 204
Windy Kansas, 279
Winnebago Indians, 95
Wintergreen berries, 9
Wintergreen leaves, 108
Witch hobble, 11, 15
Witch's-brooms, 100
Witness trees, 143
Wolf Creek Pass, 338

INDEX

Wolves, timber, 165
Wood lilies, 99
Wood thrush, 3
Woodhawks, 183
Woodland Park, Colo., 340
Woodpecker, black-backed or Arctic three-toed, 62
Woodpecker blockades starlings, 84
Woolly knees, 285
Woolly stems of mountain plants, 313
Words, inventing, 63
World of a knothole, 20
World's Columbian Exposition, 68
Wren, Bewick's, 265, 270
Wren, canyon, 202
Wren vs. flycatcher, 265

"Yankee soldiers," 38
Yeats, William Butler, 240
Yellow daisy, 336
Yellow paintbrush, 306
Yellow primrose, 213
Yellow rail, 124
Yellow-headed blackbirds, 95
Yellowstone River, 180, 184
Yellowthroat, 8
Yont, C. A., 344
"Yorick" of the veery, 13
Young, Fay H., 161
Yucca, 291
Yucca soap, 291

Zea mays, 237
Zones, life, 333

THE AMERICAN SEASONS

Edwin Way Teale's masterwork, a four volume survey of the natural history of the seasons in the United States, is now, with the present book, three fourths complete. Plans for the winter trip are being formulated. It will follow a wandering course across the continent from the southwestern corner of California to the northeastern corner of Maine.

—THE ATLANTIC.

"Edwin Way Teale is one of those fortunate few who are paid to live in the open. Writer, naturalist, and explorer of forgotten trails, he is now devoting himself to a series of four books on the American seasons. It is a magnificent assignment and Mr. Teale is just the man for it. He sees so many things we have no eyes for; all nature is his province—he is an observer in the prime, and his chronicles are diverse in color, lively and precise, his photographs magnificent. It is his special gift to take a subject of which we have a sketchy knowledge and to fill it in with fascinating detail. Mr. Teale gets the best out of people as he does out of books; he is good for wild life and particularly good for Americans who should know more about America."

—THE SATURDAY REVIEW OF LITERATURE.

"Edwin Way Teale is a writer who takes outstanding pictures and a fine, sensitive naturalist who writes excellent prose. These combined talents are one of the best things that ever happened to nature writing in America."

—THE LIVING WILDERNESS.

"Certainly the American seasons have never before had so eloquent an expositor and portrayer."

—Orville Prescott in THE NEW YORK TIMES.

(Listing the non-fiction books he thinks most likely to be read 25 years hence:) " 'North With the Spring' by Edwin Way Teale, and Mr. Teale's other nature books, all of them combinations of sound scientific observation, graceful writing and contagious enthusiasm."

NORTH WITH THE SPRING

The first volume of Mr. Teale's projected great four-volume work, "The American Seasons," takes us on a 17,000-mile journey from Florida to the Canadian border, following the advancing tide of the Spring. With many splendid photographs by the author.

—Edward Weeks in THE ATLANTIC.

"This book is Americana of the best, very human, very curious, very observant.

"But the most delightful thing about Mr. Teale is the magnitude of his interests. He is just as good with people as he is with birds, just as attentive to wild flowers as he is to the peepers."

—Robert Cushman Murphy in the NEW YORK HERALD TRIBUNE BOOK REVIEW.

"There is a candor and simplicity about the writings of Edwin Way Teale that set him apart among contemporary naturalists. He enjoys his own outdoor experiences with irrepressible gusto; and happily his written record sounds as though it had been set down for his own delectation . . . No naturalist is more scrupulous about checking facts and their interpretation."

—Joseph Henry Jackson in the SAN FRANCISCO CHRONICLE.

"They don't come any better than Mr. Teale in his field . . . In addition to making an original contribution to our country's natural history, he has written a travel book of a unique kind, a volume that any lover of nature will relish from beginning to end. It's a book that, if you let it, can keep you the best of company all winter, a book you'll honestly hate to finish."

—Van Allen Bradley in the CHICAGO DAILY NEWS.

"This will surely be ranked with the classics of American nature writing. It is an absolutely unique book, and it is packed to the boards with vivid and imaginative chapters that will entrance not only the scientist and the nature enthusiast but everyone who has ever marveled at the mystery of the changing seasons."

—BIRMINGHAM POST (England).

"One has a curious sense of the awe and majesty of Nature . . . Mr. Teale is a philosopher as well as a great naturalist, perhaps one of the greatest that America has ever produced . . . his great power of writing gives depth and colour to his book. I have read all his works, but this is far and away the best he has written. The photographs are superb."

AUTUMN ACROSS AMERICA

This second volume is a cross section of the whole continent, a 20,000-mile journey from Cape Cod to California, a wide ranging book but with particular attention paid to the four great flyways of American birds.

—Joseph Wood Krutch in THE NEW YORK HERALD TRIBUNE BOOK REVIEW.

"To accompany him on his journey is to take a twenty thousand mile 'nature walk' with a superbly competent guide and to come back with one's own eyes sharpened as well as with a thousand important or strange or puzzling facts about the things which go on in a country unequalled for the magnificence and variety of its flora and fauna as well as of its scenery."

—Sterling North in THE NEW YORK WORLD TELEGRAM and SUN.

"If you would like to expand your mind until it can grasp the sweep and beauty of our universe, take the first step now. Read 'Autumn Across America' by Edwin Way Teale."

—John Barkham in THE SATURDAY REVIEW OF LITERATURE.

"'Autumn Across America' is the best book Mr. Teale has yet written—one that captures the breadth, sweep and majesty of this prodigious country."

—THE CHRISTIAN SCIENCE MONITOR.

"In 'Autumn Across America' Mr. Teale continues the larger task of turning a naturalist's knowledge into literature. (It is) a gleaming record of a journey through the fall . . . once you have read it the world looks different."

—NATURE MAGAZINE.

". . . another Teale-rich volume . . . a moving, varied, satisfying, fact-fortified recollection in eloquent prose of an intimate and often exciting experience with outdoor America in its richest and most reflective season." (Of the four volumes:) "Mr. Teale's tetralogy will be as valid a literary form for his purpose as it is original and magnificent. Within its four-square unit we can expect to comprehend with a new awareness the whole round year and the outdoor-wealthy nation of our 48 states."

EDWIN WAY TEALE

AWARDED THE JOHN BURROUGHS MEDAL FOR DISTINGUISHED NATURE WRITING

—THE SATURDAY REVIEW OF LITERATURE.

"The American public is fortunate in having such a naturalist, writer and photographer as Edwin Way Teale. He excels in all three branches of his art, and his books . . . take their place with the best that has been done in nature writing."

—Orville Prescott in THE NEW YORK TIMES.

"A zealous scientist, a superb photographer and a sunny, pleasant writer in a great tradition, the literary-naturalist one of Fabre, Hudson and Gilbert White."

—THE SATURDAY EVENING POST.

"He has done more to bring a strange world nearer, both in the skill and uniqueness of his photographs and in the interesting observations which have gone into his books, than many men who have crossed thousands of miles of desert and ocean and have risked their lives among obscure and outlandish people."

—NEWSWEEK.

"Like Walt Whitman, Teale sees a universe in a blade of grass and can write about it in simple, convincing language. In addition he can use the camera like a pliant tool."

—BOOK-OF-THE-MONTH CLUB NEWS.

"It is amazing that books so packed with facts as Mr. Teale's can still have a leisurely pace. You can scarcely choose between the pictures and the text for charm."

—Harry Hansen in the NEW YORK WORLD-TELEGRAM.

"I treasure the Teale books, which, in my library, have their place beside William Beebe and Thoreau."

—Lewis Gannett in the NEW YORK HERALD TRIBUNE.

"Mr. Teale's books fill a special and well-loved niche."

A COMPLETE LIST OF THE BOOKS OF
EDWIN WAY TEALE

THE BOOK OF GLIDERS	1930
GRASSROOT JUNGLES	1937
THE JUNIOR BOOK OF INSECTS	1939
THE BOYS BOOK OF PHOTOGRAPHY	1939
THE GOLDEN THRONG	1940
BYWAYS TO ADVENTURE	1942
NEAR HORIZONS	1942
DUNE BOY	1943
THE LOST WOODS	1945
WALDEN (Introduction, commentary and photographs)	1946
DAYS WITHOUT TIME	1948
THE INSECT WORLD OF FABRE (Anthology)	1949
GREEN MANSIONS (Introduction, captions to illustrations)	1949
NORTH WITH THE SPRING	1951
GREEN TREASURY (Anthology)	1952
CIRCLE OF THE SEASONS	1953
THE WILDERNESS WORLD OF JOHN MUIR (Anthology)	1955
INSECT FRIENDS (Juvenile)	1955
AUTUMN ACROSS AMERICA	1956
ADVENTURES IN NATURE (Anthology)	1959
JOURNEY INTO SUMMER	1960

GRASSROOT JUNGLES

A beautifully illustrated book about American field insects, written entertainingly and with authority. Includes a special chapter on insect photography. Over 120 photographs by the author.

—Donald Culross Peattie in the NEW YORK HERALD TRIBUNE.

"There is no pleasure like standing up and praising a good book. But I am not merely standing up for Mr. Teale's 'Grassroot Jungles'! I am standing on a chair and shouting. Because it is not merely a good book; it is a splendid one. The pictures are so unusual that there is nothing like them anywhere. And Mr. Teale can narrate so that we turn the page as fast as our hands can move to find out what happens next."

—Henry Seidel Canby, BOOK-OF-THE-MONTH CLUB NEWS.

"Mr. Teale works and writes in the amiable tradition of that great lover of insect life, Fabre. His pictures are by turns exquisitely beautiful and grotesque. His text is always interesting and often eloquent. A good book."

DAYS WITHOUT TIME

Journeys into various fields of nature—the intelligence of crows, the milkweed trap, a night during bird migration at the top of the Empire State Building, etc. With 144 photographs by the author.

—NATURE MAGAZINE.

"If Teale were not so skilled a writer, such a magnificent assembly of brilliant and lucid photographs as he here presents would completely eclipse the text. It is a compliment to both his photography and his writing to note that each complements the other and neither can be said to be outclassed."

—THE SCIENTIFIC MONTHLY.

"Those who follow the genre of literature we call nature-writing are coming to regard a new book by Edwin Way Teale as an event. First of all, he is a good writer; he has something fresh to say about nature and says it directly, simply and authoritatively; he can be learned without being pedantic, and dramatic without being sensational. Second, he is a photographer par excellence."

—SOUTH BEND (IND.) TRIBUNE.

"A book with the earmarks of a classic."

THE GOLDEN THRONG

A delightfully written book on the life of the honeybee, illustrated with 85 photographs of bees in action by the author.

"This will be the Bible of the bees!"—MAURICE MAETERLINCK.

"I do not know which to admire more, his science or his art. He wins my vote on two counts."—JOHN KIERAN.

—William Beebe in THE BOOK OF NATURALISTS.

"For a play-by-play history of the world of bees read Edwin Way Teale's 'The Golden Throng,' an account doubly reinforced by a series of most excellent photographs, and unrolling in excellent English."

"The most dramatic nature book I have ever seen."
—ROGER TORY PETERSON.

NEAR HORIZONS

The story of an insect garden and of one man's adventures among mysterious neighbors within the near horizons of the insect world. Illustrated with more than 140 close-up and action photographs by the author.

—Orville Prescott in THE NEW YORK TIMES.

"It is quite possible to have no previous interest whatsoever in entomology and still succumb with practically no struggle to the fascination of this serene and mellow book. If 'Near Horizons' were only a book of educational information, it still would be of considerable value. Fortunately it is much more than that, for Mr. Teale is a skilled writer deliberately treading in the footsteps of illustrious predecessors. He illuminates his facts with the light of his own personality. Since he is good literary company, contemplative and tolerant, 'Near Horizons' is inevitably good reading."

—THE CHRISTIAN SCIENCE MONITOR.

"The seeing eye and hearing ear coupled with the fine art of story-telling (make) this one of the most attractive nature books ever published."

—SCIENTIFIC AMERICAN.

"This piece of writing should endure like those of Fabre, which it resembles."

THE LOST WOODS

Memories of the lost woods of boyhood lead the author into the adventures of a naturalist. These excursions into nature's wonderland are illustrated with more than 200 striking photographs by the author.

—LIBRARY JOURNAL.

"In pleasant leisurely prose the reader wanders along the enchanting paths of nature, ranging through the beauties of individual snowflakes; strange sights seen while flying through the heart of a cloud; the tiny insect world inhabiting a single leaf; and the grandeur of the redwood forests. The author's philosophical approach adds to one's appreciation of the satisfaction derived from studying nature in her infinite variety."

—Ludlow Griscom in THE BULLETIN OF THE MASSACHUSETTS AUDUBON SOCIETY.

" 'The Lost Woods' is . . . the most recent of a series of successful nature books. The author is a well-known literary-naturalist. He might fairly be called a modern successor to an American school which produced Thoreau, Burroughs, John Muir and Bradford Torrey."

—BOOK-OF-THE-MONTH CLUB NEWS.

"You don't have to be a naturalist to enjoy this book, but you'll wish you were one by the time you finish it."

DUNE BOY

The early years of a naturalist. A nostalgic account of the author's boyhood in the picturesque dune country of northern Indiana. Illustrated by Edward Shenton.

—THE NEW YORKER.

"A charming book. Warm with the fresh and comfortable savor of country life in the years before the first World War."

—AUDUBON MAGAZINE.

"It is hard to imagine the reader who will not enjoy this book. The writing is as sound, as homely, and as tangy as a good apple."

—THE CHICAGO DAILY NEWS.

"A host of memories crowd 'Dune Boy' and lend it something of that Tom Sawyer quality which endows stories of childhood with universal appeal."

BYWAYS TO ADVENTURE

A guide to nature hobbies, birds, animals, insects, plants, trees, weather, stars, etc. Many photographic illustrations by the author.

—BOOK-OF-THE-MONTH CLUB NEWS.
"An efficient and delightful guide to the best possible hobbies."

WALDEN
by Henry D. Thoreau

A magnificent edition of Thoreau's masterpiece, 7 x 10 inches, with introduction, interpretive comments and 142 superb photographs of scenes associated with Walden and Thoreau by Edwin Way Teale.

—THE THOREAU SOCIETY BULLETIN.
"This is *the* edition of 'Walden.' It is the edition we have dreamed of and never dared hope to see."

—Brooks Atkinson in THE NEW YORK TIMES.
"Mr. Teale is a stimulating editor. (He) has had the good sense to take Thoreau out of the classrooms and restore him to his New England pond."

THE INSECT WORLD OF
J. HENRY FABRE

The best of Fabre's writings, with a long introduction and interpretive comments by Edwin Way Teale. Photographic endpapers.

—THE NEW YORK TIMES BOOK REVIEW.
"Fabre admirers are fortunate in having in one volume so able a selection, together with Mr. Teale's charming and sympathetic introduction and incidental notes."

GREEN TREASURY

A generous overflowing anthology of the best nature writing through the ages.

—NATURAL HISTORY MAGAZINE.

" 'Green Treasury' . . . contains a great deal of writing upon which well-informed opinion will be unanimous, namely, that the samples constitute the finest writing on nature known to modern man."

CIRCLE OF THE SEASONS

The Journal of a Naturalist's Year.

—Joseph Wood Krutch in THE NEW YORK HERALD TRIBUNE BOOK REVIEW.

"None of his other books is more interesting and none, so it seems to me, is so fine as sheer writing. Moreover there is an astonishing variety and there are delightful changes of pace. Though he knows far more than Thoreau knew, he has a good deal of the same temper of mind. This is a perfect book to dip into as well as to read straight through. Something is happening on every page. Mr. Teale is one of those who know very well what Thoreau was discovering with surprise when he exclaimed: 'I had no idea there was so much going on in Heywood's meadow.' "

ADVENTURES IN NATURE

A selection of the more exciting and adventurous experiences that the author has encountered.

—Lewis Gannett in THE NEW YORK HERALD TRIBUNE BOOK REVIEW.

"Mr. Teale has been able to do just what he wanted: to wander about this country, looking at its insects, birds, wild flowers, mountains and swamps, photographing them and writing about them. It has been his pleasure and his readers' pleasure. His dozen books —from 'Grassroot Jungles' and 'Near Horizons' to 'North With the Spring' and 'Circle of the Seasons'—are contemporary nature classics. 'Adventures in Nature' is a selection from five of them . . . Anyone who reads this collection is likely to want to go to the sources for more."

DATE DUE

MR 6 '64		
MR 16 '64		
AP 7 '64		
MY 3 '64		
MY 16 '64		
JA 22 '76		
FE 5 '76		

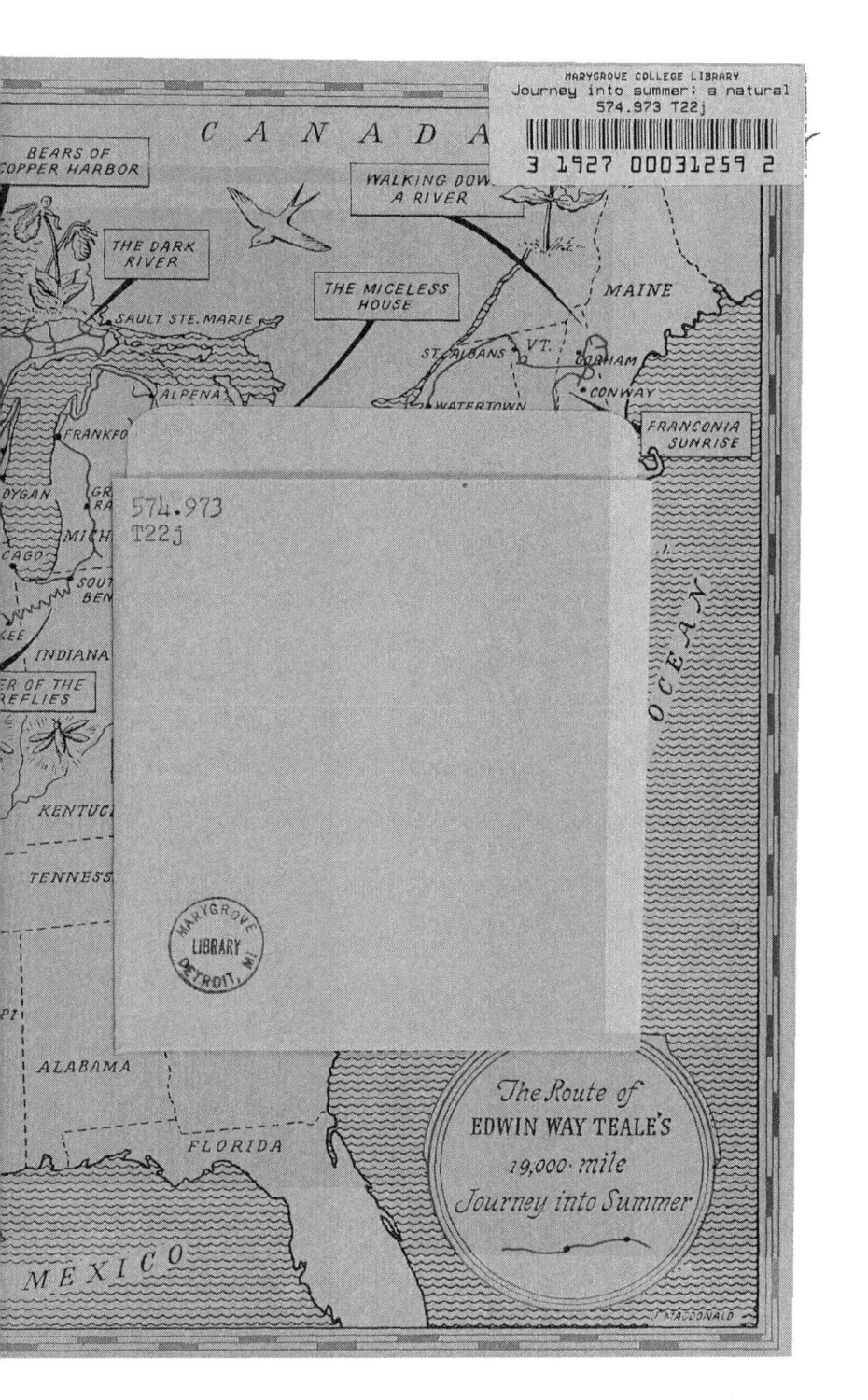

CPSIA information can be obtained
at www.ICGtesting.com
Printed in the USA
BVHW060351160223
658634BV00002B/22